The

The Final Energy Crisis

Edited by
Andrew McKillop
with
Sheila Newman

Pluto Press
LONDON • ANN ARBOR, MI

First published 2005 by Pluto Press
345 Archway Road, London N6 5AA
and 839 Greene Street, Ann Arbor, MI 48106

www.plutobooks.com

Copyright © Andrew McKillop and Sheila Newman 2005

The right of individual contributors to be identified as the authors of
this work has been asserted by them in accordance with the Copyright,
Designs and Patents Act 1988.

British Library Cataloguing in Publication Data
A catalogue record for this book is available from the British Library

ISBN 0 74532093 7 hardback
ISBN 0 74532092 9 paperback

Library of Congress Cataloging in Publication Data applied for

10 9 8 7 6 5 4 3 2

Designed and produced for Pluto Press by
Chase Publishing Services, Fortescue, Sidmouth, EX10 9QG, England
Typeset from disk by Newgen Imaging Systems (P) Ltd., Chennai, India
Printed and bound in Canada by Transcontinental Printing

Contents

Introduction

Global warming has surfaced as an issue despite being discredited, denied and rejected; sidelined by the combined forces of government, business and industry. Like fossil fuel depletion and the mass extinction of species, the topic of global warming threatens business and politics as usual, placing science in a head-on collision with the decision-making elites of the consumer democracies. Rejection of global warming by these political elites even led the drafters of the Kyoto Treaty to attempt to entice big business with cash, through the possibility of trading licenses to pollute for so-called clean development mechanism credits. Since the number-one villain in global warming is carbon dioxide, and its primary source is fossil fuel burning, the obvious solution is to burn *less* fossil fuel. Unfortunately, countless corporate and individual consumers of fossil fuels depend on them for their very sustenance and way of life. Not only business profits and political power gush forth from the oil well, but our daily bread or fast food, our pharmaceuticals, and even our clothes.

Economics, like religion, must forever be a doctrine and not a science, because it is condemned to compare oranges of today with apples of yesteryear, to compare activities of today with activities of the past, using money yardsticks whose value has substantially changed. Its real objective is in fact to create and maintain an illusion of hope. Economists will compare the mass destruction of European forests in the eighteenth and nineteenth centuries, for fuel and farmland, with the "progressive transition" to an entirely hypothetical, much-touted Hydrogen Economy, while the development of genetically modified food crops, it is claimed, will sustain us all. This is done to prove that economic growth will go on ad infinitum, the unemployed from one declining industry will be absorbed by another which grows; that any declining resource will always be compensated by another, and that human ingenuity, like human greed, knows no limits. Modern economists have the essential role of saying that economic growth is "always possible."

Peak Oil is the absolute peak in world oil production. By geological necessity, this will become an accepted fact in as little as five or six years. Under any scenario, we are entering a time when the accelerated depletion of fossil fuels must increasingly dictate events. Dramatic and even grotesque changes have been made to our planet's chemical composition. Atmospheric carbon dioxide has increased by nearly 200 per cent in less than 200 years. Climatologists and geologists who have gone public on greenhouse gases, ozone layer depletion and polar ice cap thinning have engendered ridicule from the corporate sector, often with the support of government and its economists.

As a consequence, this book is necessarily controversial. Several chapters, such as those contributed by Seppo A. Korpela and Colin J. Campbell, provide detailed, irrefutable scientific evidence of the imminent acceleration of fossil fuel depletion; while other contributors, such as Ross McCluney, Edward R.D. Goldsmith and Sheila Newman discuss the very human limits of what we currently call "sustainability." Gregson Vaux has dared to apply scientific methods demonstrating oil and natural gas depletion rates to coal. The contribution of Jacob Fisker, a particle physicist, explains in crystal clear terms why the Laws of Thermodynamics cannot be ignored in the name of economic growth theories, as they underpin and determine our very fragile existence on this planet. The late Mark Jones outlines the rivalry between China and the US, while the chapter by St. Aroman and Crouzet, discusses the extreme cost of nuclear power, which is disguised by enormous subsidies while its inherent dangers remain unaddressed. Ted Trainer has outlined several possible alternative approaches to our current lifestyle and methods. And lastly, my own contributions expose the *real* economic link between verifiable scientific fact and the political fiction manufactured by an inspired, myth-making media.

Change is inevitable. Avoiding discussion, as demonstrated by the world media's persistent praise of urban industrial civilization, glorification of consumerist ideology, and war propaganda, is not. Ignorance in the face of fossil fuel depletion, denial of planetary limits, and military responses to geophysical realities cannot vanquish anything – except our very existence. Let open, rational and scientific discussion of alternatives *and implementation and action* begin here and now, so that apocalyptic environmental catastrophe, nuclear disasters, and war might be conquered by that self-same human ingenuity which drilled the very first oil well, just over

150 years ago. Silence and political dogma can only cement our fate. Peak Oil will arrive very soon, and geological depletion will not, and cannot simply depart like an unwelcome houseguest. Time is *not* on our side, and the delaying tactics of skilled storytellers, no matter the mighty military at their command, no matter the political office they hold, will not and cannot make it so.

Part I

Depleting Energy and Bioresources: The Fossil-Fuel Key

The simplest statistical curve imaginable is an upturned, smooth-bottomed "V," called a Gaussian curve, and many physical, geological, biological, demographic and other events and relationships fit "Gaussian distributions," which are the bread and butter of statisticians because they enable *prediction*. Plotting such curves gives informed guesses of all kinds, and in the world of oil and gas (as well as minerals and bioresources) *probability curves* are the basis of all estimates, guesstimates and dispute. For oil, the most classic Gaussian curve is called the Hubbert Curve, of which you will see several in the carefully researched and detailed chapter by Korpela. First plotted by M. King Hubbert in the 1950s, this curve maps the success rate of drilling measured by production and accumulated production, to find when the absolute peak of production will likely occur, in the rounded upper part of the curve. After that, analysis based on various "profiles" for success rates and remaining oil reserves allows various scenarios to be graphically shown for the coming downside leg of the curve.

Just as no one will ever repeal the Second Law of Thermodynamics, almost nobody can seriously challenge M. King Hubbert, although heroic efforts are made. As Korpela and Campbell clearly state, worldwide oil and gas reserves *will* be depleted or exhausted in a profile similar to Hubbert's Gaussian curve, at a level of probability a 95 per cent or above. Consumers of *any petroleum product or of natural gas* around 2035 will really be a dying species, somewhat like the entire class of reptiles and amphibians in Europe, whose species numbers are decreasing about 4 per cent every year. This will also be the *likely annual falloff* or "average decay rate" for oil production and use through about 2008–35. Preceding this, there may well be sharper annual declines (perhaps well above 6 per cent) for several years, slowing as the curve flows outward and downward. Fast decay rates will then return near the end. By about 2035, oil production will be down by around 75 per cent from today's levels of about 78 million barrels/day, and through unstoppable demographic growth the per capita consumption of oil, oil products and natural gas will have fallen by about

95 per cent, and at least 65 per cent for gas, compared with today. Compressed into less time than from when the first Concord thundered into the stratosphere, burning up to nine liters per second of kerosene, to 2003, the wipeout of cheap oil and gas will be dramatically rapid.

There are many statistical whorls and frills to Gaussian curve drawing, and deriving short-sectional parts of curves (for example around the peak, or at the very beginning and end) from an overall lifetime curve. Notably, the *rates* of change of the plotted variables are much more violent in these sections, and for world oil, whose overall "useful lifetime" will be about 1860–2040, the large changes in rates of production – and therefore consumption – were from the late 1960s to early 1980s on the upward part of the curve. For the downside leg of the curve the corresponding periods will be about 2008–18. This has more than a few implications for us all. In the 1960s the only problem with oil supplies was finding ways to use more, faster. Machines like the Concord and B-52 bombers were invented, and proudly utilized, to destroy oil resources and also human life. World oil demand increased regularly at 7 per cent per year, and from 1970–73 even attained 11 per cent per year. On the downside, from about 2010 this experience – in pure statistical probability terms – should be reversed, with "minus" signs as the former growth rates become decay rates. Decades ago, this was warned in various ways, from the Club of Rome's "Limits to Growth" study, and "Blueprint for Survival" by the *Ecologist* editorial team, to the first Stockholm conference on the environment in 1971–73. The initiator of "Blueprint for Survival," Edward R.D. Goldsmith, contributes to this section.

In the early 1970s, at least one quite large oil region – the North Sea – remained to be developed, major technology improvements in prospecting and production were coming, and in the early 1980s there was a ferocious economic recession, producing a sudden slowdown in oil demand growth. The over-zealous and overconfident presentations of Hubbert-type curves in the early 1970s, showing a full stop to world oil supplies on exactly January 1, 2001, did much to ruin the appeal and credibility of his approach. Any Gaussian distribution curve actually shows that almost anything never runs out completely; there are simply phases and periods of *rate change* for the increase or decrease of some variable, event or thing – in this case, world oil supplies. However, the Hubbert curve, when applied to more reasonable estimates of remaining world oil reserves, gives us clear warning: in graphic terms, it was and is the shape of things to come.

Apart from the creation of the "limit denial" industry, little or nothing has happened so far by way of real responses to having to do without, or using something else. This particularly concerns the significant area of food production, which Goldsmith addresses head-on. In some ways, we "eat oil." The world's biggest consumer countries for nitrogen and phosphorus-based fertilizers correlate to their populations: China, India, and Indonesia all use more than the US; in China's case, about six times more. Fertilizers are effectively impossible to produce without oil, gas or coal. Back-of-envelope calculations, and those detailed in other chapters, show that if we suppose there are *no* fossil fuels available for fertilizer production (we can also imagine 50 per cent and then 75 per cent less), there will be at least a 50 to 60 per cent fall in the world's carrying capacity for human beings. In other words, a not very well fed world of about 3 billion population, last attained in 1962–63, has more than doubled in the 40 years of cheap oil and energy that followed. Under any hypothesis, fertilizers will become more expensive because of coming peaks for oil and gas, and farming will become more difficult because of the cheap energy interval and its negative impacts on soil quality, water supplies, pathogen numbers, and farming practices. Thus, the crying wolf of world energy supplies in the early 1970s, echoed in coming world famines, was essentially right. Just as with any new fashion collection, the timing has to be right. In the early 1970s the last bulge of world oil discoveries had just terminated, and the last bulge of world gas discoveries had not begun; cheap oil and gas were going to produce the massive quantities of fertilizers needed to feed another, and perhaps the last, doubling of world population.

Today more than ever, the "limit denial" industry is in full flood: well financed and basking in the acclaim of right-thinking media manipulators. Any subject either directly covered by this book, or partly addressed (such as climate change) has its perfect antidote on the Internet, in the press, and on television and radio. Climate change negationist sites on the Internet – some of them directly financed by the US government – number more than 50,000. Curiously, this hardly exists for oil and gas. Denial of depletion is an academic and conference circuit specialty, with imaginative redrawing of Hubbert curves (to show Hubbert got everything wrong even when he was right!) and ever more impressive *reserve growth* of oil reserves "discovered" in paper and digital archives.

The very biggest players – Exxon-Mobil and BP-Amoco, reigning numbers one and two in the world oil and gas business (despite their own production capacity constantly decreasing: see Chapter 2) – play things in a totally, even schizophrenically different fashion. ExxonMobil does not deny depletion at all. It states very clearly that world oil is depleting at a rate of about 3.25–3.5 million barrels per day lost in capacity each year, although it calls this "both economic and geological depletion." ExxonMobil adds that to cover depletion *and* satisfy growing demand, the world must spend about US$250 billion per year, every year, on exploration and development over 2003–15. The real rate of spending today is far behind that. We can suggest economic depletion simply means that producing oil in certain areas – as with heavy oil, shale oil and very deep offshore oil, gets too expensive as costs mount, and timelines to produce smaller or more difficult reserves get even longer. In other words, unless prices move up sharply that production capacity can be lost: a "downspike" of prices causes a shutdown of production. After a while, production becomes too expensive even to start.

BP-Amoco takes a vastly different tack. This Anglo-American oil corporation already suffered a mild nervous breakdown in 2001, when it announced itself "Beyond Petroleum," but later denied anything of the kind. It now calls itself "Beyond Petroleum" only for its solar energy marketing activities (and in Russia could be nicknamed "Buying Politicians"). Since 2003, BP has revealed, in a flurry of academic and theoretical studies, that its painstaking, well-funded research shows world oil demand will soon cease to grow altogether. By 2007 or 2008, demand growth will stop dead in its tracks – whether through rising prices, or because the Chinese and Indians decide to return to bicycles and ox carts, BP does not say. In fact, BP is saying in an oblique, easily deniable way that Peak Oil could come as soon as 2008; hence world oil demand will stop growing by 2008.

Everybody thinks they know what "energy" means, but before saying so they should try Fisker's chapter on the meaning of thermodynamics. The subject is not at all limpidly clear and without ambiguity – but the very name explains what energy is about. Thermal energy and motion, or "dynamics," are related. Without the sun there would be no life on this planet. Our existence as rather fuzzy-edged, evaporating human beings (millions of atoms disappear from your body each millisecond), having a low density of about 0.96, or less than water, depends on multibillion degree temperature nucleosynthesis in the sun. Our planet is so far away that it would be $-5°C$ if it were not for the atmosphere which serves us in many ways. Things run our way, down that thermodynamic sink. This is why life exists and works so well on our planet, despite periodic cataclysms – for example, the "Permian die-off" and the saga of dinosaurs which is part and parcel of, and physically linked to the gas, but not the oil, in your homes, cars and other fuel burning equipment. Fisker spells out the laws of thermodynamics which, unlike human laws, can neither be broken nor bent. The bottom line is very real and significant for us: there are only so many "stock" forms and *quantities* of energy, while "flow" forms continue as long as the sun goes on shining but are *low intensity*, relative to accumulated "stock" energy. Large "stock" reserves like fossil oil, gas and coal – or uranium – are finite and depleting the second they start to be used. Their formation took so many millions of years that we can regard them as one-time or one-shot resources. Whenever depletion results in more energy being needed to extract, process and supply energy, the process and its form are doomed. This is coming rapidly for oil, and somewhat more slowly for world natural gas supplies. Plenty of coal remains, while uranium supplies are very energy-intensive to produce, and above all create open-ended security and pollution risks.

Conversion to renewable energy sources will come, and facilitating this enormous transition – rather than denying its causes and the imperious need for it – should be the task of all persons, including political, business, theological and educational elites. Currently we have a denial industry, and the confusion of violent price movements on the so-called "free market." The timeline for this non-system to collapse is very short – perhaps not even ten years. New and rediscovered old ways of using less energy better will be found and applied. As the 1980s politicians used to chirp, as do their clones of today: "You have no alternative."

I
Prediction of World Peak Oil Production

Seppo A. Korpela

Those who read articles on oil in the daily papers often encounter the statement: *with present rate consumption oil will last 40 years*. This number is obtained by dividing the reported reserves by annual production. Since the reported reserves are in round numbers, 1,000 billion barrels (billion barrels can be written as giga-barrels, or Gb), and 25Gb is used for annual consumption, the outcome is 40 years.

These are soothing words, as 40 years is beyond the lifespan of most readers. It takes an alert reader to see that oil production will not stay flat for 40 years and then suddenly drop to zero. Rather, it will rise to a peak, after which mankind is faced with an era of declining production. Thus it is clear that *peak production* is the most important event regarding our future reliance on petroleum, and the media and newspapers would do us all a service by reporting just how close this might be.

The subject of predicting peak oil production occupied the mind of M. King Hubbert during most of his professional life. After obtaining his graduate degrees from the University of Chicago, and serving seven years as a geology instructor at Columbia University, he moved to Shell Research Laboratories and became its director, later to join the United States Geological Survey (USGS). While at Shell Research he delivered a paper at the meeting of the American Petroleum Institute in 1956, where he predicted that oil production in the US would peak in about 1966 if the lower figure of 150Gb of total oil, was accurate; or 1971 if the higher estimate of 200 billion barrels was to hold. These two estimates for *ultimate production* were judged by oil geologists to be reasonable at that time. In 1956 Hubbert had not yet developed the mathematical method he later used for estimating US ultimate production, but used a graphical procedure to come to his conclusions.

By 1962 he had developed a procedure to draw what is now called *Hubbert's Curve* for the future oil production in the lower 48 US states. Since his method relies on production history, and no oil had yet been produced from Alaska or the deepwater domain, he excluded them from consideration. By this time, he had settled on 170 billion barrels as an estimate for the ultimate production in the lower 48 states, a figure he did not alter in his later study of 1972, although he lowered it to 165 billion barrels in 1980.[1] When US oil production attained its all-time peak in 1970, Hubbert's methods proved to have predictive value. Alaskan oil, and more recent oil from the deeper parts of the Gulf of Mexico, will alter ultimate oil production for the US, but the lower 48 figure is not likely to change much.

With the passing of the Hubbert Peak for US oil production in 1970, and his estimate that the world peak will appear around 2000, the next generation of petroleum geologists took up the analysis and prediction of Peak Oil. Most notable of these are Colin Campbell, Jean Laherrère, L.F. Ivanhoe, Walter Youngquist, and Kenneth Deffeyes. In the March 1998 issue of the *Scientific American*, Campbell and Laherrère published a joint article, "The End of Cheap Oil,"[2] in which they masterfully laid out the foundation for the reading public to understand the situation regarding world oil. Campbell's monograph,[3] which appeared the same year, is a longer exposition of the subject, with exhaustive graphs and tables to support his warning that the era of cheap oil is nearly gone. He has alerted his professional colleagues to this dilemma in the *Oil & Gas Journal*, the leading oil industry trade journal.[4]

Campbell divides oil-producing countries into three groups: those that are past their peak production; those that are near the peak but have not quite reached it; and the so-called "swing producers," all located in the Persian Gulf region. Swing producers are called upon to supply that shortfall as other producing countries, one by one, will pass their production peaks and – as their production falls and domestic consumption grows – cease to be oil exporters. Having access to an industry database, Campbell is able to track production and depletion patterns in the various large petroleum basins of the world. Using this data, he makes yearly assessments for major producers, and reports the results in the Newsletter of the Association for the Study of Peak Oil (ASPO).[5]

Laherrère has refined and extended Hubbert's methods. As a consultant to the oil industry, he has access to the same industry database as Campbell does. This enables him to track discovery trends accurately and use them in his analysis. These show that the peak for world oil discovery took place in the early 1960s, and that there remain approximately 200 billion barrels still to be found, which *at current production rates* would provide about eight years' supply. This is a mere 10 per cent of the world's initial or ultimate oil endowment, meaning that *90 per cent has already been discovered.* His graphs, which show how cumulative production lags accurately backdated oil discovery by some 38 years, ought to convince anyone of the nearness of the coming oil production peak.[6]

Of the other notable analysts in this field, Ivanhoe is a director of the M. King Hubbert Center for Petroleum Supply Studies at the Colorado School of Mines and contributes a quarterly newsletter. In this he discusses production and consumption patterns of the major consuming and producing countries by geographical groupings.[7] Duncan, an electrical engineer, has developed an heuristic model into which he supplies inputs based on new information, which he collects from petroleum geologists around the world, as well as economic factors influencing demand.[8] He collaborates with Youngquist, who is a petroleum geologist, and whose book *Geodestinies* contains a wealth of information on all aspects of mineral resources.[9] Finally Bakhtiari, the senior analyst at the National Iranian Oil Company, contributes articles to industry publications, such as the *Oil & Gas Journal*, on reserves in the Middle East.[10] He is skeptical that Persian Gulf countries can increase production at the rate suggested by the International Energy Agency, which would demand an increase of net oil exports from the Middle East OPEC producers from about 19 million barrels/day (Mbd) in 1997, to 46.7Mbd in 2020.

An excellent account of Hubbert's methods is given in the recent book by Deffeyes, who provides not only a primer in petroleum geology but also interesting historical comments on the life of Hubbert at the Shell Research Laboratory.[11] This presents and discusses two mathematical models for predicting peak oil production. As new data become available each year, constants in each model can be re-estimated, and the accuracy of prediction thereby improved. For this reason, when applied to the oil production history and outlook for the US, which is in its late phase of decay or decline, the models are in excellent agreement. With the benefit of US experience, the models are next used to show that world conventional oil production is essentially at its peak today, with production of all liquid hydrocarbons peaking by 2010, assuming that demand remains flat or increases from here on. How well these models predict decline for world oil is of lesser importance, as political events, which the models obviously cannot predict, will undoubtedly influence the decline. By then the Final Energy Crisis will be upon us, and we will likely have more important things to do than to fine-tune mathematical models for the decay rate.

In discussing oil production there is one more aspect that needs to be clarified: namely, what to count as oil. In the production of oil, some gaseous components condense as their pressure is reduced on entering from the well. This is called lease condensate and, as it is liquid and comes from oil fields, it is often counted as oil. Similarly, as a part of natural gas production, the heavier hydrocarbons condense and are counted as part of the liquid stream. Similarly, refinery gains arising from separation of the hydrocarbons of differing molecular weights add to the volume of liquid produced. These are the two main reasons why world oil production today is reported as being either 24 billion barrels/year, or the 16 per cent higher figure of 28 billion barrels/year. The data used here is for oil and condensate for the US, and oil only for the world. Heavy oil, tar sands, deepwater and polar oil are more expensive to produce and, although partly in the stream today, are likely to be more important after world conventional oil has peaked.

HUBBERT'S METHOD AND LOGISTIC EQUATION

The method used by Hubbert to predict peak production for the lower 48 US states is based on the *logistic equation*. The same equation was used by Verhulst to make calculations of human population growth in 1838. Only in 1980, by which time the US production peak was past, did Hubbert give a full account of his methods.[12]

To understand the mathematical basis of Hubbert's method requires some knowledge of elementary calculus. Let Q denote the cumulative amount of oil that has been produced in some large oil province from the beginning of production to the present. The logistic equation states that the rate of increase of cumulative production, Q', which can be taken to be the *annual production*, is given by the equation

$$Q' = aQ(1 - Q/Q_0)$$

Here, a is a parameter controlling the sharpness of the peak and Q_0 is the *ultimate production*. This is to say, the amount of oil that has been produced when the oil province is finally abandoned. Those familiar with calculus will recognize that Q' denotes the derivative of Q. Inspection of the right hand side of this equation reveals a parabola, which increases from zero when Q is zero to a maximum, then drops to zero again at $Q = Q_0$. The maximum value is $aQ_0/4$, at the midpoint of ultimate production.

Before taking up the solution to the logistic equation, consider the early years of production. Then Q is much smaller than Q_0 and the logistic equation can be written in the approximate form

$$Q' = aQ$$

This is the same equation used for calculating interest payments and sums, where Q represents the principal, which when multiplied by the interest rate a gives the yearly interest earned, Q'. For continuously compounded interest the familiar exponential growth formula is obtained.

For oil production cumulative production increases as

$$Q = Q_i \exp[a(t - t_i)]$$

Here, Q_i is the cumulative production at time t_i, and exp denotes the exponential function, with t the present year. The rapidity of growth is determined by a, which is called the *intrinsic growth rate*.

A different and worthwhile view of the logistic equation is obtained by recasting it in terms of how much of the ultimate *remains to be produced*. If this is denoted by $Q_r = Q_0 - Q$, then toward the end of oil production the annual production can be determined from

$$Q_r' = -aQ_r$$

From this, the remaining reserve declines or decays as follows:

$$Q_r = Q_i \exp[-a(t - t_i)]$$

Thus, the remaining oil decreases exponentially, and the parameter a can now be called the *intrinsic decay rate*. Hubbert recognized that early in the production cycle cumulative production increases exponentially, and at the end it decreases exponentially. On this basis he drew by hand the possible production curve such that the area under his curve equaled the ultimate.

The ultimate production can be shown to follow the formula

$$Q = Q_0/(1+\exp[-a(t - t_m)])$$

where t_m is the year of peak production, at which time cumulative production is $Q = Q_0/2$.

The three parameters in the solution are a, Q_0 and t_m. The most straightforward approach to estimate two of them is to recast the logistic equation into the form

$$Q'/Q = a - aQ/Q_0$$

and to interpret it as a straight line from coordinates showing Q on the horizontal axis and Q'/Q on the vertical. The slope of this line is negative and has the value a/Q_0. It intersects the vertical axis at a, and the horizontal at Q_0. Hence from this plot estimates can be made for both a and Q_0, from the actual production history. The value of t_m can be determined from the known production record, which shows that at time t_i the cumulative production is Q_i. The only thing left is to decide which year and thus which pair of values to choose for t_i and Q_i.

PRODUCTION CURVES FOR THE US

The data to determine the US production are available from the US Energy Information Administration (EIA) website,[13] and from Campbell's monograph.[14] The actual data are plotted in Figure I.1.1 in the manner outlined above, with every tenth year identified by a filled circle. Inspection of this graph shows that Hubbert's 1956 estimate for ultimate production – in other words the *initial oil endowment* of lower-48 US oil – of between 150 and 200 billion barrels, is reasonable. The data in the early phase of a production cycle are subject to fluctuations, as yearly production is divided by cumulative production, which remains small in the early phase. In 1962 Hubbert gave an estimate of 170 billion barrels, the best one could do with the data at hand. He reported an intrinsic growth factor of 0.067, when time is given in years, during this pre-peak phase.

During the early years, independent analysis of discovery trends gave estimates of varying reliability for the ultimate reserve. With any ultimate figure at hand, the straight line can be forced to cross the horizontal axis at this value, and a least squares fit can then be used to determine the best value for the intrinsic growth factor. Of course, at that time the mathematical basis for calculating discovery trends was not yet known, and estimates were based on the hunches of petroleum geologists on the amount of oil in any given region.

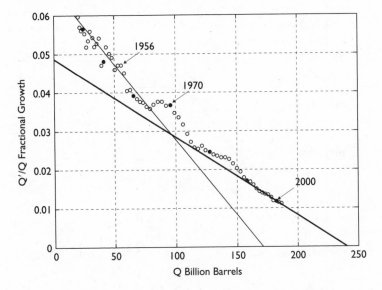

Figure I.1.1   A plot to estimate the parameters for US oil production, where Q is the cumulative production and Q' is the annual production.

Today, a least squares fit through the last ten years gives the heavy line shown in the graph and an estimate of $Q_0 = 240$ billion barrels for the ultimate recoverable reserve of the US, and $a = 0.049$ for the *intrinsic decay rate*. The viewpoint now changes to decay, as the peak is past. These values put the theoretical production peak at 1977.

Slightly different results are obtained if greater number pairs of data are used in the fit. If the years since 1942 are used peak production occurs in 1975, and the original endowment becomes 222 billion barrels. The official estimate of the USGS is 362 billion barrels.[15] That the estimate by USGS differs from the Hubbert type of analysis must mean the USGS team has no confidence in Hubbert's methods, or requires an optimistic viewpoint on national reserves. This is not new, as Hubbert himself was faced during the 1960s with ever-increasing estimates for the ultimate. In the 1960s these optimistic estimates grew to 600 billion barrels in the hands of McKelvey, the Assistant Chief Geologist of the USGS.[16] Hubbert was a member of USGS staff from 1964 to 1976 and, when reading his 1982 report, one can surmise how the ever-increasing reserve estimates must have given him impetus to put his method for estimating the reserve base on a very solid scientific footing. Even if the latest estimate of the

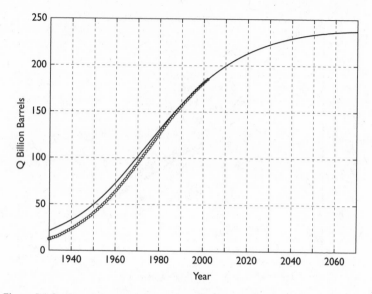

Figure I.1.2 Cumulative production of oil in the US, where the solid line is the theoretical prediction and open circles are the actual production data.

USGS has declined substantially, one is at a loss to find a scientific explanation for the high value of 362 billion barrels still being published. USGS does not project actual past discovery and production, but makes abstract geological assessments with rather arbitrary and subjective probability rankings, and publishes the mean value of reserves. This has been subjected to criticism by Campbell.[17] The cumulative production can now be plotted (see Figure I.1.2). The cumulative amount produced in the US to the end of 2002 is 186 billion barrels. Since the published reserves are 22 billion barrels, this leaves 32 billion barrels to a category of *reserve growth* in existing fields and yet-to-find fields, if 240 billion barrels is used as the figure for the ultimate production.[18]

It is remarkable how well the general trend of the theoretical graph follows the actual data. One might object that the excellent fit is primarily owing to data from a long historical record being used to predict the parameters in the model, and that it may not have had this predictive capability before these data were available. But, had the data from 1958–66 been used in 1967 to predict the peak, the result would have shown year 1976 to have been the peak year for US production. Of course, the actual peak had already taken place in

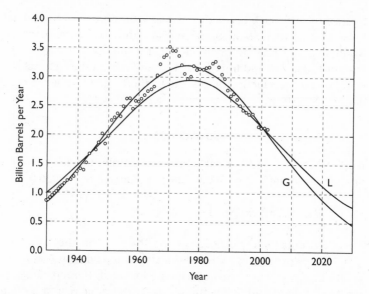

Figure I.1.3 Annual oil production in the US. The graph labeled L is obtained from the theory based on the logistic equation, while that labeled G is based on the normal or Gaussian distribution of statistics.

1970, six years earlier, because the theoretical peak had moved to a later date, owing to Alaskan and deepwater Gulf of Mexico oil finds.

Annual production is shown in Figure I.1.3. The curve labeled L is obtained using the logistic equation. The actual data show the oil peak in 1970. The subsequent decline to a bottom in 1975 was the result of conservation efforts after the first oil crisis. The rising trend to year 1985 was caused by a drilling boom in the Gulf of Mexico and completion of the Alaska pipeline, directly due to impacts of the 405 per cent nominal price rise for crude oil between 1973 and 1981. Deepwater oil production in the Gulf of Mexico may slow the rate of US depletion for the next few years, but once this production starts to diminish the decline rate will of course accelerate, and is likely to return to the trend rate of decline, as it did after Alaskan oil was largely consumed.

The key to Hubbert's prediction was his recognition that oil production *must* follow its discovery pattern. Through his study of US discoveries, which peaked during the early 1930s, he was able by 1956 to make his bold predictions concerning US oil, before he had developed the mathematical basis on which to apply rigor to his forecasting. Although his prediction was dismissed, it has turned out to be quite accurate. Many today will also dismiss the prediction of world oil peak being imminent, despite the improvement in reserve reporting, analysis and forecasting techniques, and the fact that the world peak of production is closer to us than when Hubbert carried out his 1956 study.

GAUSSIAN MODEL

Bartlett has used the normal (Gaussian) distribution to model oil production in the United States. The Gaussian curve is given by

$$Q' = [Q_0/\sigma\sqrt{(2\pi)}] \exp[-(t_m - t)^2/2\sigma^2]$$

Bartlett obtains $Q_0 = 222.2$ billion barrels for the ultimate, $t_m = 1975.6$ for the peak year, and a standard deviation $\sigma = 27.56$ years. After two more years of data are included, these parameters are 220.7 billion barrels for the ultimate, the peak year at 1975.6 (i.e., July 1975), and 27.52 years for the standard deviation. The Gaussian curve is shown in Figure I.1.3 as the curve labeled G, whereas that based on the logistic equation is labeled L.

The reason the Gaussian model represents the data better at the peak is that it minimizes the error between the model and the data over the entire set of yearly production values. As has been discussed, the logistic model is *forward-looking* and enables prediction of the decline from now on. How well the logistic equation predicts future production depends on the amount of oil that will be produced from the deepwater region of the Gulf of Mexico. The Minerals Management Service of the US Department of Interior estimates that oil production from these deeper parts of the Gulf of Mexico will rise to somewhere between 2.0 and 2.47 million barrels/day in the year 2006, from 1.5 million barrels/day in 2001. They have estimated that the entire basin will contain 71 billion barrels of oil, of which 56 billion are hoped-for or yet-to-find.[19] If all 56 billion barrels of this hoped-for oil should be found, the ultimate endowment of the US will exceed the current estimate of about 240 billion barrels. However, a part of this deepwater oil already shows up in the production data, and using the forward projecting approach for estimating the ultimate reserve by the logistic equation will correct data as each year goes by. The extra oil, should it turn up, will provide some cushion, and would certainly go further after the world peak is past, when awareness will force thrifty habits on a larger part of the population.

WORLD OIL PEAK

The arrival of world oil peak is a much more serious issue than the US peak 30 years ago. It will usher in the final energy crisis. The following section employs the above methods for predicting US oil production decay to the question of predicting the date and volume of world peak oil production.

Logistic Model

To obtain the intrinsic growth or decay rates and the ultimate for world production, data are plotted in Figure I.1.4 using the logistic equation. It also contains the US data for purposes of comparison; the scale along the top refers to US production. The trend lines are remarkably similar, and on this basis one might estimate how world peak production will evolve in time.

A least squares fit through the last ten years gives the estimates $a = 0.0456$ and ultimate production $Q_0 = 2,212$ billion barrels. These put the world peak six years away, in early 2009. As world

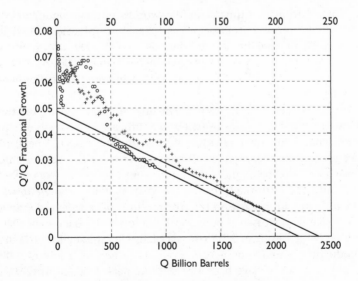

Figure I.1.4   A plot to estimate parameters for world oil production. Open circles are for World oil; crosses refer to US oil.

yearly consumption at the 2002 rate of production is 25 billion barrels, each 100 billion barrels added to the world's ultimate reserve *delays* the production peak by about two years, and the ultimate reserve lifetime by four years.

Year 2000 is, so far, the year of peak production. A 2,400Gb ultimate would delay the theoretical peak until year 2012. Since production at the peak is quite flat, the model gives a production rate in 2012 only 6.2 per cent larger than in 2000. This is in sharp contrast to the demand-driven projection of the IEA, which calls for an increase of about 23 per cent in production from now to 2012.[20]

The estimate of 2,400Gb is about the most generous that the data will permit, and would require many more deepwater discoveries to be made and brought on stream, and at record speed. It might also be accomplished by US intervention in Iraq and the discovery of petroleum that has so far been overlooked. Such a hope may be misplaced, as half of Iraqi reserves lie in just three fields. This chapter was written on the very day UN Resolution 1441 was approved (November 9, 2002), when a possible oil war was still inching its way forward. However, mankind has fought resource wars through its history. As this book asks, why should the present age be any different, despite the growing number of states with nuclear weapons?

A more realistic estimate for the world's ultimate endowment, including yet-to-discover reserves, is 2,200Gb, as this calculation has shown. This places the peak production year at 2009, with production at about 25.2Gb/year, 2.8 per cent higher than in 2000 and 4.7 per cent higher than today. This means that compound annual growth can be only 0.77 per cent. Conversely, annual world oil demand growth has been close to 1.8 per cent since the early 1990s, and the New Industrial Countries of East and Southeast Asia have typically shown national growth rates for oil consumption of 3.5 to 5 per cent per year, China's growth in oil consumption for the year ending November 2002 being more than 6 per cent.

Finally, the fit through the last ten years may in fact *overestimate* the ultimate, for the reason that the recent past experienced strong oil demand. Increased production has moved the data somewhat higher, which leads to 2,400Gb as a plausible estimate for ultimate production. Should the actual turn out to be much lower, 2000 could turn out to have been the absolute peak production year in world oil. Production has been down since late 2000, owing to recession and recent price increases resulting from a Venezuelan

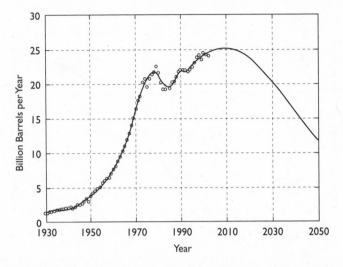

Figure I.1.5 Annual projected world oil production. Theoretical peak production is predicted to take place in 2009.

shutdown and fears of war. This seems to demonstrate how little spare capacity there actually is.

The rise and fall of world annual production is given in Figure I.1.5 for the reference case, with peak production in 2009 at a slightly higher rate than in 2000. It is for these reasons that Campbell, in his yearly assessment of world oil, places the next ten years on a production plateau for conventional oil, with rising production from deepwater reserves until 2010. In fact, if demand slows everywhere the world production profile might easily look like that for the US, with a fall in production for the next few years followed by a slight secondary peak early next decade. This will again leave 2000 as the absolute peak.

Since non-conventional oil, including that obtained not only from deeper parts of the world's sedimentary basins but also from the polar regions, is more costly to produce, prices and production during the plateau will be determined by an interplay of political events and economic factors, matters that are amply discussed elsewhere in this book. Under any hypothesis, however, there is little prospect of cheap and abundant oil remaining a fixed or "given" part of the world economic situation.

The curve through the data until 1992 is drawn to aid the eye, using actual data; the curve elements for the last ten years and future

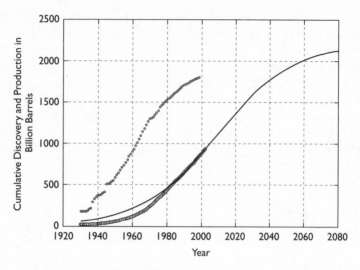

Figure I.1.6 World cumulative discovery and cumulative production.

production are from the logistic model. In fact, only the annual production for the past ten years and cumulative production calculated at the start date of 1992 are needed to project future production. This ought to put to rest the objection that the logistic model is "unsuited" to such forecasting because it produces a symmetrical production history.[21] The shape of the early production history is irrelevant, and it does not matter how it fluctuated. The fall and subsequent rise during the 1980s are likewise irrelevant for determining future production. In addition, re-examination of the production profile for the US shows that its broad features are quite symmetrical, and the reason for drawing the entire theoretical curve for US production in Figure I.1.3 was to show this symmetry. One might still quibble and say that peak production for the US was in 1970, while the logistic model delays it to nearly 1977. If Alaskan oil is excluded, the theoretical peak is closer to the actual.

Cumulative production for the world is shown in Figure I.1.6. The left-most curve is an approximate reproduction of the data from Laherrère's paper on cumulative discovery.[22] He obtains it by back-dating any reported reserve additions to existing fields, so that they represent oil in the original find, underestimated at first but revised using new knowledge of the reservoir, as the production progresses and field size are better delineated. To carry this out, each revision must be assigned to the proper field, which of course means that field-by-field data must be available. Since these data are difficult to obtain, the reader is referred to the original article by Laherrère, for the same figure with his discussion. The discovery history can also be modeled by Hubbert's methods, and since it produces a similar pattern roughly 40 years earlier, using either the US or world data, it is unquestionably straightforward to predict how the world's oil production history will evolve. The deviation in the production history for the early periods shown here is a result of using the intrinsic decay factor, instead of the intrinsic growth factor, to fit the early periods.

Figure I.1.6, as well as the cumulative production model, shows the strength of Hubbert's methods for predicting future oil production. The objection that the model is not solidly based on geological fact, and hence not reliable, is misplaced. The claim that the model is too simple to represent the multitude of factors that must drive oil consumption is also without merit. Whereas it is true that the model involves only two parameters, there are situations in other fields in which complicated models can be reduced by rigorous mathematical methods to simple forms, when and if the parameters in them happen to have magnitudes that allow such simplification. In addition, as has been discussed, the parameters in Hubbert's model can be, and ought to be, recalculated each year. When the model is used with knowledge of backdated reserves, it enables increasingly better predictions to be made each year.

Laherrère has improved on the Hubbert analysis by recognizing that many countries have more than one discovery cycle – as for example when offshore reserves were opened late – each of which can be modeled with a logistic equation; the production for the country then equates to the sum of their Hubbert curves. Still, Hubbert's methods work best by applying them to large regions, as factors that may have political origin in various parts of the world tend to cancel each other out.

Gaussian Model

Bartlett fitted the Gaussian model to world data and found that the best fit yields $Q_0 = 1,115Gb$ for the ultimate. He mentions that the reason for this stems from the data lacking a prolonged downturn.[23] Although this is true, the fundamental issue is that this problem – of non-linear regression analysis – needs to be handled differently from the way that he proceeds. When a non-linear regression

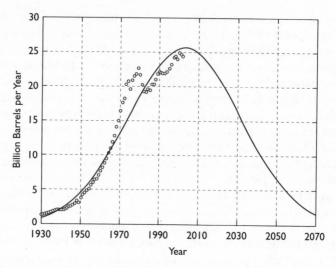

Figure I.1.7   Annual oil production for the world and its future projection. Peak production according to the Gaussian model took place in October 2002.

analysis is done, it gives $Q_0 = 1,833$Gb for the ultimate, $t_m = 2002$ for the peak year, and $\sigma = 28.6$ years. The ultimate is lower, and the peak one year earlier, than was calculated by Deffeyes.[24] Whereas the backdated reserves give a time lag of 38 years between discovery and production, Deffeyes follows Hubbert and, using published reserve data, develops a time lag of 21 years for the delay between discovery and production. By fitting a Gaussian curve through the discovery, Deffeyes establishes the ultimate endowment as 2,120Gb for world oil, with world peak production occurring in 2004. Although the accuracy of the data does not warrant this level of precision, the calculations carried out here point to October 8, 2002, as the day of ultimate peak oil production, humorists adding that it was at 2:47 p.m. A graphical presentation is given in Figure I.1.7. Since the entire data set is used, the early phase of the production history, with its sharper rise, pulls the Gaussian curve to the left, and this persists through the growth phase of the last dozen years. Hence the peak is *earlier* than given by a forward-looking logistic curve.

The theoretical point of whether annual production should follow a Gaussian curve rather than a Hubbert profile, by virtue of the "central limit theorem," is an interesting issue in statistics. However, in the world as it is, where political events caused a dip in consumption and a boom in exploration through the 1980s, no predictive model can forecast such real-world effects on the demand side. Whether oil production in a world so loved by economists, where all humans make rational decisions based only on their perceived utility, could or should follow a strict Gaussian curve without deviation is also an idle question. This did not happen in the past, and future events are likely to distort trends even further. Economists tend to look at price signals to project future scarcity. If actual oil production peaks by the end of this decade, while price signals – that is, large price rises – only appear several years later, those price signals will certainly have been inferior in predicting scarcity than the logistic model discussed here. The "consensus view" is that as prices rise, currently uneconomical oil will be produced, and reserves will increase through presently uneconomic resources being produced and consumed.

Such an argument ignores the energy cost of production, and the question of Energy Return On Energy Invested, or EROI, as discussed elsewhere in this book. When the energy for production is used in increasing quantities from the output itself, reducing net energy, clearly there comes a point at which the net energy delivered becomes zero. This will happen *long before* the world's difficult-to-produce (or "frontier" or "non-conventional") oil has been completely produced and consumed. Furthermore, oil is a remarkably useful fuel, as it exists in liquid form at atmospheric temperature and pressure, giving a high energy per unit of volume. Economists would have us believe that an equally good substitute will emerge to replace it, but no one is yet ready to mention what that might be. Hydrogen is a non-starter, as it is an "energy sink" – whereas oil and natural gas presented their ready-to-use, versatile energy as a gift from nature.

CONCLUSIONS

Two mathematical methods to predict the world oil peak have been discussed in this chapter. Hubbert's method, which is based on the logistic equation, is shown to be superior to the Gaussian model. The Gaussian model makes use of all the production data, and must necessarily produce a symmetric fit. With the fit constructed so that the error has been minimized over the entire data set, the resulting graphs look the best and thus have some appeal. However, there is no compelling reason to favor this method. The method based on the logistic equation, developed by Hubbert and refined by Laherrère, is dynamic, forward-looking and good at predicting future trends. When used with the discovery trend its results provide the best method to predict the coming peak in oil production and the onset of the era of oil scarcity.

This chapter was written in November 2002. This paragraph has been added in early March 2003. In December 2002 a general strike commenced in Venezuela, and during its aftermath it is likely that perhaps 250,000 barrels per day of production has been permanently destroyed. The imminent war in Iraq has caused the price of oil to reach a high of US$40 a barrel. The growing scarcity of natural gas in North America has increased its price threefold over one year. These two factors have made it difficult to switch from natural gas to fuel oil in industry and power production. Should the war in Iraq destroy her oil infrastructure, the final energy crisis will have commenced by the time this book is in the reader's hands.

2

The Assessment and Importance
of Oil Depletion

Colin J. Campbell

Oil provides some 40 per cent of the world's energy needs and as much as 90 per cent of its transport fuel. It also has a critical role in agriculture, which provides food for the world's population of 6 billion people. It is however a finite commodity, having been formed in the geological past, which means that it is subject to depletion. Given that it is of such great importance to the modern world, it is indeed surprising that more attention has not been given to determining the status of depletion.

There are several possible explanations for this strange state of affairs. First, it is counter-intuitive. The weekly trip to the filling station is such a normal part of daily life that most people see a continued supply of oil as being as much a part of nature as are the rivers that flow from the mountains to the sea. Second, depletion is strangely foreign to classical economics, which depict Man as the master of his environment under ineluctable laws of supply and demand. Never before have resource constraints of such a critical commodity begun to appear without sign of a better substitute or market signals. The reason for the absence of early market warning is due to expropriations that have obscured the natural trends that would otherwise have alerted us to growing shortages and rising costs. Tax by both consuming and producing countries has furthermore distorted the position. A related issue is a blind faith in technology, as epitomized by the dictum "the scientists will think of something." Unfortunately, if they do, they will simply deplete the remaining oil faster. Third is the denial and obfuscation by the oil industry, which is in a position to understand the situation but finds itself the victim of an investment community driven by imagery and a very short-term view of the future. The industry is itself subject to internal vested interests, represented for example by the explorers, whose careers are not served by pointing out the natural limits. The oil companies can accordingly be excused for choosing their words with such extreme care. Fourth is the nature of so-called democratic government that finds it easier to attract votes by reacting to crises than by anticipating them, especially where unpopular or even draconian responses are called for. Fifth are possible conspiracies by countries already dependent on rising oil imports, which seek to secure access to supply and hold prices down by any means at their disposal.

There may be other factors at work too, but these five elements offer a range of possible explanations for why the subject is so clouded by mystery and disbelief. In strictly technical terms, there is nothing particularly difficult in assessing the size of an oilfield or in extrapolating the discovery trend to indicate what remains to be found in the future. I will try here to lift the veils, to provide a fair statement of the true position.

BACKGROUND

Oil from seepages has been known since biblical times, being used for example as mortar in Babylon, but the modern oil industry had its roots in the nineteenth century, when it commenced drilling for oil on the shores of the Caspian and in Pennsylvania. The technology was not new, as wells had been drilled earlier to tap salt brines, needed to preserve meat in the days before refrigeration. The science of petroleum geology evolved rapidly to provide a technical basis for exploration. It was soon appreciated that oil resulted from the decomposition of microscopic organisms and that, once formed, it could be trapped in certain geological structures, which could be identified and mapped.

Attention turned to the Middle East during the early years of the last century, prompted first by seepages in the Zagros Foothills of Iraq and Iran, but later by the less obvious prospects beneath the sands of Kuwait and Saudi Arabia, which were identified only with the help of core-drilling and seismic surveys. The prolific oil lands of the United States, Mexico, Venezuela and Indonesia were also opened up in parallel. Exploration expanded throughout the world, so that most of the onshore oil basins and many of the giant fields within them had been identified prior to World War II.

The demand for oil grew rapidly from the economic boom that followed the war, prompting further exploration, which led to important finds in the Soviet Union and in Africa, as well as the opening of offshore drilling, which was greatly facilitated by the development of the semi-submersible rig in 1962. Floating on submerged pontoons beneath the wave base, it brought routine drilling to the continental shelves of the world.

From the earliest days, the oil industry has been characterized by "boom and bust" cycles, for the simple reason that, once found, oil flows from the ground at great pressure, which contrasts with the painfully slow extraction of coal and minerals by pick and shovel. Prolific production from new finds flooded the market and depressed the price, which in turn inhibited new exploration until the early wells began to run dry, when the cycle was repeated by new discovery in new areas. It became evident that some control of the open market was called for to avoid these damaging fluctuations. The US was the first to apply it, using the Texas Railroad Commission to maintain price by rationing production. Its example was followed by the creation of the Organization of Petroleum Exporting Countries (OPEC) in 1960, which sought to perform the same much needed function on a world scale.

The wealth from oil, known as black gold, became legendary. The industry grew to be the largest in the world, and many of the producing countries began to rely heavily on oil. As their economies and populations grew, so did their appetite for oil revenues, which in turn led to expropriations, prompted also by the belief that they were not being fairly compensated for the depletion of their natural resources. The Soviet Union was the first to move, expropriating the assets of the foreign companies in 1928, followed ten years later by Mexico. The trend accelerated after the war, starting in Iran in 1951, and extending during the 1970s to the other main producers – Iraq, Kuwait, Saudi Arabia, Libya, Venezuela and Algeria. But it did not quite deliver the anticipated benefits. The international companies had been previously able to take the revenues paid to the foreign producing countries as a charge against home country tax. The producing countries lost this hidden subsidy following expropriation when they had to face the raw pressures of the open market. Prices became volatile as a consequence.

In the period 1947–49 Britain, which had administered the territory of Palestine as a protectorate, surrendered to terrorist pressures. Massive Jewish immigration led to conflicts and the occupation of new lands, forcing the indigenous Arabs into refugee camps. Tensions mounted in 1973, causing certain sympathetic Arab oil producers to restrict exports to the US and the Netherlands, which were perceived to side with Israel in the conflict. Although it was a short-lived restriction of only a few months, oil prices rose five-fold in what became known as the First Oil Shock, which plunged the world into recession, curbing oil demand. It was, in turn, followed five years later by a second price shock, occasioned by panic buying following the fall of the Shah of Iran, when oil prices soared to the equivalent of US$80–US$100 a barrel, in today's money.

These two oil price shocks, while themselves transitory and politically motivated, demonstrated the degree to which the world had become dependent on cheap oil. That in turn prompted several contemporaneous studies of depletion, epitomized by the well-known report "The Limits to Growth," which in fact, like "Blueprint for Survival" (from *The Ecologist*) was published just before the first Oil Shock.

The resources themselves were far from running out, but it was realized that production would eventually reach a peak and decline. This obvious conclusion was not exactly new; in 1956 M. King Hubbert had already drawn a simple bell-curve showing that US production would peak in 1971, at the midpoint of depletion. He could readily estimate the total endowment below the curve, as discovery had peaked 40 years before the study. Since oil has to be found before it can be produced, it is obvious that production has to mirror discovery, after a time-lag.

But the warning signals were both ignored and misrepresented, as new production from Alaska and the offshore, including particularly the prolific North Sea, flooded the world with cheap oil. Having lost their principal sources of supply through expropriation, the international companies concentrated on these new areas, which they did control, and worked flat-out. In Europe, socialist governments, which could have managed their countries' resources in the national interest through state companies, were replaced by doctrinaire free marketers, epitomized by Mrs. Thatcher, who encouraged the rapid depletion of Britain's oil and gas. The North Sea has now peaked, and is declining at about 6 per cent a year, meaning that production will have roughly halved in ten years' time (see Figure I.2.1).

In 1981 the rate of discovery began to fall short of consumption, despite a surge of tax-driven drilling, and the deficit has grown ever since (see Figure I.2.2). Although the international companies continued to speak optimistically about their future growth, their actions told a different story. By the end of the century, the so-called Seven Sisters, which had dominated the world of oil for so long, had been reduced to just four: Shell in splendid isolation, Exxon-Mobil, BP-Amoco-Arco, and Chevron-Texaco, with a second tier in Europe

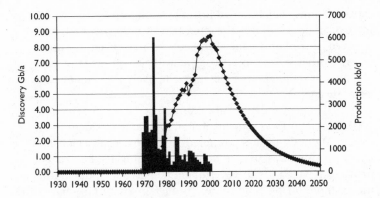

Figure I.2.1   Oil discovery and production in the North Sea.

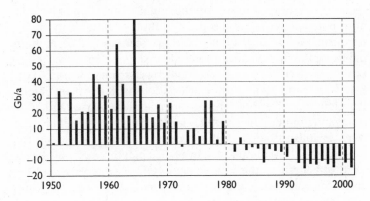

Figure I.2.2   Discovery–consumption gap.

comprising of Total, Elf and Fina. The major companies began to merge, downsize and shed staff. One changed its logo to a sunburst and claimed that its initials stood for "Beyond Petroleum," as a very oblique reference to the depletion of its principal asset. The so-called independent companies, too, were disappearing through merger and acquisition, as was the contracting business, on which they all depend.

The production of conventional crude oil outside the five main Middle East producers reached a peak in 1997, but falling demand from an Asian recession, combined with increasing Russian exports made possible by the weak ruble, led to an anomalous fall in oil prices. That was, in turn, followed by the reappearance of supply capacity limits, which caused prices to triple – that is, a 200 per cent increase in price – during the latter part of 1999. The high price was believed, in certain financial milieux, to have been a cause of the recession that started in 2000, which weakened demand, reducing pressure on oil prices in a vicious circle likely to be repeated in the years ahead.

The last chapter of this unfolding saga came on September 11, 2001. The action was soon attributed to a Saudi dissident living in a cave in Afghanistan, and prompted the US to declare a global war on terrorism. Afghanistan was bombed, toppling its Taliban government, with which the US government had no difficulties in dealing – for example on pipeline routes through Afghanistan, and within months before September 11. During Soviet occupation, US governments had actively supported and financed Osama bin Laden. Israel took the opportunity to step up the brutal suppression of its indigenous population, claiming common cause with the US in a war on terrorism. A brief unattributed anthrax scare in the US mobilized public opinion, bringing unprecedented popularity to President Bush, who declared several oil producing countries to be an "Axis of Evil" – the term "axis" having associations with the perpetrators of the Holocaust. He threatened to invade oil-rich Iraq, and started stockpiling military fuel supplies for the purpose. US military bases were established around the Caspian oilfields, while in Latin America the US was implicated in a failed plot to overthrow the president of Venezuela, the strong man of OPEC. Observers can be forgiven for concluding that the new foreign policy of the United States has a hidden oil agenda.

What may follow from this chain of events is impossible to predict, but it is at least on the cards that popular outrage in the Middle East, prompted by US military intervention, may lead to the fall of the Saudi government, giving the US the pretext to take the Arabian oilfields by force. The US has long explicitly declared that it regards access to foreign oil as a vital national interest, justifying military intervention where necessary. As its own domestic production continues to decline without hope of reprieve, its need for foreign oil becomes ever more desperate. Whereas in the past military intervention may have been contemplated as a reaction to politically motivated interruptions to supply, now it faces the inevitable consequences of depletion and conflict, with other countries also seeking a share of what is left. Much of that lies in the Middle East, thanks to circumstances in the Jurassic period over which no politician can exercise control.

MISLEADING OIL REPORTING

The main reasons why this subject is not better understood are the ambiguous definitions and unreliable reporting practices of the industry.

Conventional and Non-Conventional Oil

Oil is oil from the standpoint of the motorist filling his tank, who does not much care from whence it comes. But the analyst of depletion needs to identify the different categories, because each has its own costs, characteristics and extraction rates, and hence can contribute differently to peak production. The term *conventional* is widely used to describe the traditional sources, which have contributed most oil produced to date, and which will dominate all supply far into the future.

There are, in addition, *non-conventional* sources, which will be increasingly important when conventional oil declines after peak, but there is no standard definition of the boundary. Here, the following categories are treated as non-conventional, and are described in greater detail in a later section.

Oil from coal and "shale" (actually immature source-rock)
Bitumen
Extra-Heavy Oil (density $<10°$ API)
Heavy Oil (density $10–17.5°$ API)
Polar Oil and Gas
Deepwater Oil and Gas (>500m water depth)

Liquids that condense naturally from the gas-caps of oilfields are included with crude oil, but the liquids extracted from gas by processing are treated separately.

Production and Supply Reporting

Measuring production is simply a matter of reading the meter, but national statistics are confused by the inconsistent treatment of natural gas liquids, war loss (which is production at least in a technical sense) and frontier changes. Supply is not the same as production, but includes stock change and refinery gains. The reporting of gas production is still more confused, referring variously to raw gas or marketed gas after the removal of inert gases such as nitrogen and carbon dioxide, which are often present, with the differing treatment of flared and re-injected gas adding to the uncertainty.

Reserve Reporting

The practices of reserve reporting evolved early, being much influenced by the environment of the old onshore fields of the US, which were characterized by a highly fragmented ownership. Being onshore and close to market, the wells could be placed on production as soon as they had been completed. To prevent fraud, the Securities and Exchange Commission (SEC) introduced strict rules whereby owners could treat as *proved* for financial purposes only the reserves "behind pipe," meaning those to be drained by existing wells. As the fields were drilled up, the reported reserves naturally grew. In practice, the reserves of the old fields were mainly estimated by extrapolating the decline rates of the wells. The highly fragmented ownership meant that there was little interest in, or indeed possibility of, making field-wide reserve estimates. The companies did however recognize additional *probable* and *possible reserves* that did not qualify for proved financial status.

Different conditions obtained overseas and in offshore areas, where it was normal for fields to be developed as single entities by one or more companies acting as a group. They were more interested in what the field as a whole would produce over its full life, especially offshore, where they had to design appropriate facilities in advance of production. They were still lumbered with the SEC rules, which required the reporting of proved reserves, although in practice they reported better estimates of what the fields would deliver over their full lives. There is naturally a degree of latitude in estimating future production. For a variety of commercial reasons, it was found expedient to report ultra-conservative estimates of discovery, which consequently grew over time, delivering an attractive impression of gradually appreciating assets to the stock market, and serving to reduce tax in countries operating a depletion allowance. Many of the large North Sea fields, for example, were initially underreported by about one-third. This luxury is not however available to the more recent small fields, with a short life and high economic threshold. They in fact sometimes give disappointing results, yielding negative reserve growth.

A further confusion has arisen from the application of probability theory, in which reserves are equated against differing subjective probability rankings. Under this system, *proved reserves (1P)* are commonly equated with a 95 per cent probability, whereas *proved & probable & possible reserves (3P)* are held to have a 5 per cent probability. *Mean*, *median* and *mode* values are then computed. This system appeals to the scientifically inclined, but in practice adds to the confusion. In plain language, *proved reserves* relate to the current status of development, whereas *proved & probable reserves* (or '*Mean*' under the probability system) are estimates of what the field as a whole is expected to produce over the rest of its life. The probability range is unnecessarily wide, as engineers with modern methods can make good estimates. Why should anybody be interested in an assessment having no more than a 5 per cent probability of being correct, and a subjective one at that?

The Dating of Reserves Revisions

For financial purposes, reserve revisions are reported on the date that they are made, but this gives a misleading impression of the discovery trend. Year-on-year comparison of national reserves, with subtraction of the intervening production, gives the impression that more is being found than is the case, which has misled many analysts working with data in the public domain.

To determine a valid discovery trend, it is necessary first to make sure that the reported production and proved & probable reserves relate to the same categories of oil; and second to backdate any revisions to the discovery of respective fields. In practice, this cannot be done without access to the industry database to identify the details, and its cost puts it out of range for most analysts.

Several OPEC countries announced colossal overnight reserve increases in the late 1980s, when they were vying for quota based on reserves. While some upward revision was called for, as the earlier numbers were too conservative, having been inherited from the private companies before they were expropriated, the revisions had to be backdated to the discovery of the fields containing them, some of which had been found up to 50 years earlier.

Dating the reserves is as important, if not more so, than estimating the amounts. The explanation for the revisions is another highly important matter to grasp. The industry, not wishing to admit to poor reporting practices, has found it expedient to attribute revisions to technological progress when they were in fact mainly a reporting phenomenon. This in turn carries the danger of unjustifiably extrapolating reserve growth into the future on the assumption of an inexorable march of technological progress. No one disputes the progress to-date, but its main impact has been to hold production higher for longer, which makes good economic sense but accelerates depletion.

Data Sources

Two trade journals, the *Oil & Gas Journal* and *World Oil*, have compiled information on production and reserves for many years on the basis of a questionnaire sent out to governments and others. Many of the reports remain implausibly unchanged for years on end, simply because the country concerned has failed to update its estimate, despite production. There are also substantial discrepancies between the two data sets despite the fact that they are compiled in a similar fashion.

A third source is the *BP Statistical Review of World Oil*, which is the most misleading of all, because many analysts wrongly assume that the reported oil reserves have at least the tacit blessing of a competent and knowledgeable oil company in a position to assess their validity. In fact, BP simply reproduces the *Oil & Gas Journal* oil reserve data, save in one or two specific cases.

These public sources contain information very different from that in the industry's own database, which is compiled on a field-by-field basis directly from the companies' own records. This itself contains certain anomalies, and seems to be deteriorating in quality as it faces the increasingly difficult challenge of compiling information from the proliferation of small companies and ever less reliable state information. Particular difficulties are faced in interpreting data from the former Soviet Union, which operated its own system of reserve classification that tended to ignore economic constraints.

In short, although there are no particular technical difficulties in estimating the size of an oilfield, especially with the advantage of modern technology, the reporting of production and reserves remains highly unreliable. In these circumstances it is well to confirm, wherever possible, the estimates of individual fields by extrapolating the decline, which plots as a straight line on a graph relating annual to cumulative production (see Figure I.2.3).

ESTIMATING FUTURE DISCOVERY

In earlier years, it was difficult to identify the precise source of the oil that found its way into oilfields, but a geochemical breakthrough in the 1980s resolved the issue. Isotopic examinations showed that oil was derived from algae (and similar micro-organisms), whereas gas came from vegetal material, as well as deeply buried oil, which had been broken down into gas by high temperatures. This knowledge led in turn to the realization that the bulk of the world's oil

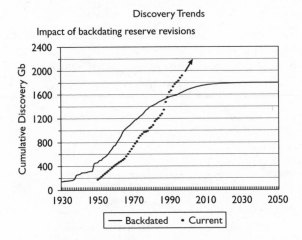

Figure I.2.3   Annual vs. cumulative production.

came from no more than a few epochs of extreme global warming, when prolific algal growths effectively poisoned the seas and lakes. Petroleum geology made great advances, making it possible to map the world's producing belts once the critical data had been gathered from seismic surveys and preliminary boreholes. The world has now been so extensively explored that virtually all the productive belts have been identified, save perhaps in certain polar and deepwater regions, which are here treated as non-conventional, partly for that very reason.

Estimating the future discovery of an established basin is a straightforward task, achieved by extrapolating the discovery trend with a so-called *creaming curve*, which plots cumulative discovery against cumulative *wildcats* (exploration boreholes), and by studying field size distributions with a *parabolic fractal*. The larger fields are generally found first, for the simple reason that they are difficult to miss, being followed in turn by progressively smaller finds (see Figures I.2.4 and I.2.5).

A FLAWED STUDY BY THE UNITED STATES GEOLOGICAL SURVEY

(See also Chapter 17)
The USGS started evaluating the world's oil resources following the oil shocks of the 1970s, and under its previous director put out

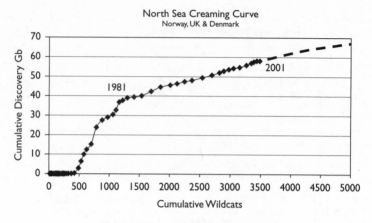

Figure I.2.4   North sea creaming curve.

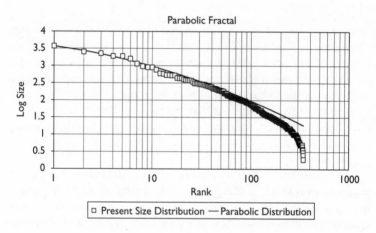

Figure I.2.5   Parabolic fractal.

sound evaluations that were published at successive World Petroleum Congresses. A departure was issued in 2000, which greatly exaggerated the scope for new discovery and the "growth" of existing reserves. It is worth briefly commenting on this flawed study because it has misled several foreign governments and agencies, including the International Energy Agency.

The study started by usefully identifying all the prospective basins of the world. It did not extrapolate past discovery with the methods outlined above, but relied on abstract geological assessment, subject to a range of subjective probability rankings. Thus for example, in the case of an un-drilled basin in East Greenland, it determined that there was a 95 per cent chance (F_{95}) of it containing more than zero, namely at least one barrel, and a 5 per cent chance (F_5) of it containing more than 112Gb, from which a mean value of 47 billion was computed.

In reality, the 5 per cent chance cases cannot be other than wild guesses that could as well give half or double the true value, yet they influenced the computation of mean values, which were summed to give the world total. The indicated amounts related to discovery over a 30-year period starting in 1995. While there may be some abstract scientific merit to the study, it says little about what will actually be found in the real world, as is well confirmed by the results to-date. The mean estimates imply an average discovery of 25Gb a year, when so far the actual average has been only 10Gb, which is doubly damning because above-average results are to be expected during the early years, as the larger fields are normally found first.

The estimates of "reserve growth" are equally flawed, being based on the experience of the old onshore US fields, which, as discussed above, are not remotely representative of the offshore or overseas. To its credit, the USGS did express serious reservations about the estimations in the accompanying text. While the study itself speaks of academic inexperience and ineptitude, political overtones were introduced when the USGS issued a press release of the unfinished study on the eve of a critical OPEC meeting, and by the fact that it goes to great lengths to publicize the study at conferences around the world and by direct interventions with foreign governments and agencies.

It is, at the same time, curious to find a member of the USGS team publishing an impressive poster that depicts the imminent peak of oil production, termed the great Roll-Over, with a text speaking of a rough ride if the world does not wake up to the reality of its predicament.

ESTIMATING FUTURE PRODUCTION

If we had reliable production and reserve data, and it is a very big *if*, it would be a fairly straightforward task to forecast future production using one or more of the following statistical techniques:

1. *Simple depletion* The simplest, and in some ways perhaps the best model, is to divide the world into three groups:
   - countries past their depletion midpoint, where production is expected to decline at the current depletion rate (annual production as a percentage of total future production);
   - countries that have not yet reached their midpoint, whose production is set to continue to rise until midpoint, before declining at the then depletion rate;
   - swing countries, comprising the five major Middle East producers (Abu Dhabi, Iran, Iraq, Kuwait and Saudi Arabia), which make up the difference between world demand under various scenarios and what the other countries can produce under the model. The current base case scenario is that conventional production will be on average flat as a result of alternating price shocks and consequential recessions until 2010, when the swing producers can in practice no longer offset the declines elsewhere, and world production commences its terminal decline at the then depletion rate.

   This method is used in the ASPO Statistical Review. It has the advantage of recognizing demand impacts, not easily covered in the strictly statistical methods described below.

2. *Hubbert models* Production in an unfettered environment can be modeled with a simple Hubbert bell-curve based on an estimate of ultimate recovery, or with multiple curves reflecting different cycles of discovery and corresponding production. In world terms, a simple Hubbert curve, built on the indicated size of the resource (1,900Gb), shows a peak in 1995 at 40 million barrels/day, but was not realized because the oil shocks of the 1970s curbed demand, giving a lower and later peak (see Figure I.2.6).

3. *Discovery–production correlation* Since production has to mirror earlier discovery, future production can be modeled by superimposing the production trend on the past discovery trend with a time shift, as demonstrated by Laherrère.

4. *Rate plots* There is a mathematical procedure that converts a bell curve into a straight line, achieved by plotting annual production as a percentage of cumulative production on one axis, against cumulative production on the other. The straight line can be readily extrapolated, as explained by Deffeyes.

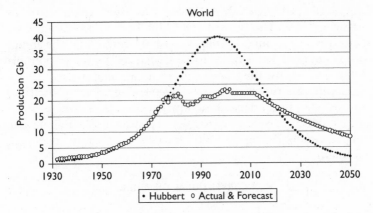

Figure I.2.6   Hubbert curve.

GAS

Gas is more difficult to evaluate than oil. Some occurs in discrete deposits, known as dry gas, being mainly derived from deeply buried coals, but most in the gas-caps of oilfields, known as associated gas. About 80 per cent of the gas in a reservoir is recoverable, compared with only about 40 per cent for oil. Liquid hydrocarbons, known as *condensate*, condense naturally from gas on being brought to the surface, and more may be extracted by processing, both forming important resources for the future.

Gas depletes very differently from oil due to its higher mobility. An uncontrolled well would deplete a gas deposit quickly, and production is normally deliberately capped far below the natural capacity, commonly by the simple expedient of pipeline pressure. Accordingly, production normally follows a long plateau, with most fluctuation being seasonal. Gas prices generally fall as the investment costs are written off, which in turn attracts new customers. Production continues along the set plateau for a long time, but when the inbuilt spare capacity has been drawn down, it comes to an abrupt end without many market signals. Whereas oil trade is global, the gas market is regional, built around the hubs of North America, Europe and the Far East. The US faces a crippling shortage of gas as it reaches the end of its plateau of production, and will rely increasingly on imports from Canada and the Arctic. Europe is better placed, being able to draw on supplies from North Africa, Russia,

Central Asia and eventually the Middle East, after its own North Sea production ends. The Far East and China have to depend on local sources, as well as imports from Russia and Central Asia. The abrupt end of gas production needs to be recognized by the responsible authorities, because the market delivers no warning signals.

Modeling gas supply, with its hidden in-built spare capacity, is difficult, as so much depends on infrastructures and markets. Here, global production is expected to rise to a plateau of 170 trillion cubic feet per annum, lasting from 2015 to 2040, but the forecast is most uncertain.

NON-CONVENTIONAL OIL AND GAS

Heavy Oils

Oil may be extracted from coal by the use of the Fischer Tropf process, invented in Germany during the war, and it may be retorted from immature oil source rocks, termed "oil shales." Much interest was shown in the latter method after the oil shocks of the 1970s, but all projects came to naught. The residue is a fine, toxic powder carrying environmental hazards and costs, and the net energy return is very poor.

Conventional oil migrated to the margins of the basins of western Canada and eastern Venezuela in substantial quantities, where it was weathered and attacked by bacteria. The light fractions were removed, leaving behind bitumen (defined by viscosity) grading into extra-heavy oil (defined by density). In Canada, the so-called tar-sands containing the bitumen are mined at the surface after the removal of up to 75 meters of overburden. The ore, for that is what it is, is centrifuged and processed in plants fueled by cheap stranded gas to yield a light, high-quality synthetic oil. In Venezuela the deposits lie at 500 to 1,500 meter depths and are produced with the help of steam injection from closely spaced wells. The extra-heavy oil grades into heavy oil, which is here arbitrarily defined as that denser that 17.5° API. There are other deposits around the world, but those of Canada and Venezuela are the most important.

The resources are enormous, but the extraction rate is low and costly. No doubt production will be stepped up from the current level of about 2 million barrels/day after the peak of conventional oil, but it is difficult to imagine it exceeding about 5 million barrels/day by 2020, despite superhuman effort and every financial incentive. It is also worth remembering that the deposits are not homogeneous: even a small addition in the thickness of overburden adds greatly to the cost of tar-sand extraction. Processing also uses fuel, which will become increasingly expensive once the stranded gas deposits, currently used, have been exhausted.

Deepwater Oil and Gas

In earlier years, the deepwater domain was considered too far from land to contain oil reservoirs and source rocks. But recent exploration has identified certain areas in divergent plate settings where Cretaceous rifts yield source rocks, and where turbidity currents comparable with submarine avalanches brought in sands to form reservoirs, especially where winnowed by long shore currents. These special conditions appear to be restricted to the Gulf of Mexico and the margins of the South Atlantic. Deltas elsewhere may locally extend into deepwater, but lacking underlying prolific source-rock, any petroliferous potential they might have will rely upon whatever source-rocks occur within the delta itself, which are likely to be gas-prone. Present evidence points to a total endowment of about 65Gb. If all goes well, production may peak at around 8 million barrels/day within a year or two of 2010. It is axiomatic that no one would look for oil under these extreme conditions if there were anywhere else easier left.

Polar Oil and Gas

Antarctica appears to have very limited geological prospects, and is in any case closed to exploration by agreement. The Arctic regions are more promising, although large vertical movements of the crust under the weight of fluctuating ice caps in the geological past have tended to depress the source-rocks into the gas window. Alaska is an exception, but appears to be concentrated habitat with most of its oil in the giant Prudhoe Bay field, which has been in decline since 1989. There are also substantial oil deposits in the Siberian Arctic, here tentatively estimated at 30Gb, with production reaching a peak of about 5.5 million barrels/day by 2020. The gas reserves throughout the Arctic are likely to be very large indeed, but extraction will be slow and costly in this extreme environment.

Non-Conventional Gases

Coalbed methane, derived from coal deposits, is an important non-conventional gas already supplying about 6 per cent of US needs. More can be expected from the other coal-bearing regions of the

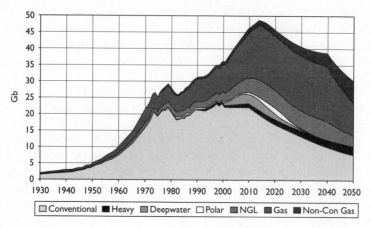

Figure I.2.7   Depletion of all hydrocarbons.

world. Another useful source is gas extracted from fractures in hydrocarbon source-rocks, known as "tight gas." Much attention has been given to gas hydrates, which, it has been wrongly claimed, form large deposits in deepwater and polar regions. In reality, the methane occurs in disseminated granules and laminae, which are unlikely to be producible.

Figure I.2.7 depicts the depletion of all hydrocarbons as modeled herein. It will be noted that the production of conventional oil is expected to be about flat until 2010, due to alternating price shocks and consequential recessions dampening demand. The peak of all liquids also comes around 2010, with gas following about 15 years later. It means that the production of all liquids need not fall below present levels for about 20 years, assuming that the deepwater and polar oil, and natural gas liquids, come in as expected. In the unlikely event that sustained economic growth could be restored, demand would rise accordingly, advancing the peak and steepening the ensuing decline.

WORLD REGIONAL ASSESSMENTS

United States

Discovery in the US peaked in 1930, followed 40 years later by the corresponding peak in production. Alaska provided a secondary cycle, but was insufficient to reverse the decline, and the new

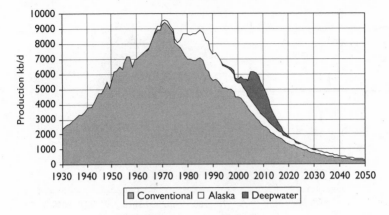

Figure I.2.8   US production.

deepwater Gulf of Mexico offers a third (see Figure I.2.8). It is doubted if the Alaska Natural Wildlife Refuge (ANWR) area, which is closed for environmental reasons after the drilling of one very confidential borehole, would make any material difference if opened. US oil imports already run at about US$130 billion a year and are set to rise unless the government can somehow introduce draconian policies to cut demand. Its gas supply is even more critical, as already discussed. It is hard to avoid the conclusion that this looming energy crisis will spell the end of the American dream and US global economic hegemony, even if the country goes down with all guns blazing.

Russia

Russian discovery peaked in the 1960s, followed in 1987 by a peak in production at just over 11 million barrels/day. Production, which fell precipitately on the collapse of the Soviet government, is now set to increase to a second slightly lower peak around 2010, in part bringing in what would have already been produced but for the interruption. In addition there may be substantial production of non-conventional oil from the Arctic (see Figure I.2.9). Russia's gas deposits are very large indeed, with reserves amounting to about one third of the world's total, which will see increasingly high demand from Europe, China and the Orient.

Russia is experiencing an epoch of hyper-capitalism, with the emergence of various oil barons who could put Mr. Rockefeller to

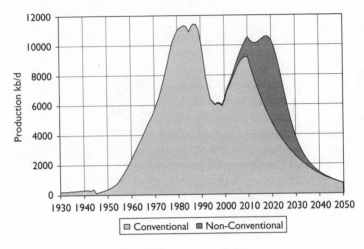

Figure I.2.9   Russian production.

shame. At the present time they are bent on exporting at the maximum rate possible to earn foreign exchange, and are able to undercut world prices thanks to the devalued ruble, which holds down their operating costs. It is entirely possible that Russia may take over from the Middle East the role of swing oil producer during the next decade or so, and it is already building a dominant position in Eastern Hemisphere gas markets. This confers great geopolitical strength, and responsibility both for the country itself and the world as a whole. Those conscious of the iron grip of depletion might conclude that Russia's national interest would be well served by producing at a low rate so as to make the resource last as long as possible. In commercial terms, it might find advantage in providing its own manufacturers with reliable and even cheap energy, rather than subsidizing its competitors.

The Caspian Chimera

(See also, Chapter 5)

The Caspian is the oldest oil province in the world, where the Tsars established an oil monopoly even before Col. Drake drilled his famous well in Pennsylvania. The activities have been concentrated on the shores of the Caspian, especially around the early oil centre of Baku in Azerbaijan, as the Soviets did have the need to exploit offshore drilling. The area became of great interest to the West at the fall of the Communists, and some wildly exaggerated hopes that it would replace the Middle East were aired, based yet again on the flawed studies of the USGS. The first problem was to decide who owned it: if it was deemed to be a lake, international law required that its resources be jointly exploited by the contiguous countries, a solution favored by Iran and Russia; but if it was deemed a sea, it would be divided up by median lines, as in the North Sea – a solution favored by Kazakhstan. The Western companies, however, moved in without waiting for this little matter to be resolved.

In geological terms, offshore Caspian reserves can be divided into four provinces. In the south lies a deep gas-prone tertiary basin, which has yielded the Shah Deniz Field, operated by BP. To the north is a narrow belt, forming the proto-delta of the Volga, which extends from Baku to Turkmenistan, becoming gas-prone in that direction. The results to date have been disappointing, causing Exxon-Mobil to withdraw. Future production may not exceed about 10Gb from known and yet-to-find fields. Next comes a modest Jurassic trend that extends out of Kazakhstan, offering perhaps another 5Gb. Lastly, in the far north, comes the southern limit of the prolific Pre-Caspian Basin, most of which lies onshore. Interest here was stimulated by the Tengiz Field, found in 1978 by the Soviets, with about 6Gb of high sulfur oil in a Carboniferous reef at a depth of over 4,000 meters, which is now being developed by Chevron. A huge structure, called Kashagan, was identified in the adjoining waters of the Caspian. Had it been full of oil, it might have justified the exaggerated early claims, but three wells have now been drilled at enormous cost, suggesting that it is made up of several discrete reefs, with a potential in the 10–15Gb range. At all events the results were sufficiently disappointing to cause BP and Statoil to withdraw from the venture. So, a sanguine estimate suggests that no more than about 30Gb are likely to be produced from the offshore Caspian, which is equivalent to approximately half the North Sea's resource. This is indeed useful and valuable production, but it is unlikely to have any particular impact on global supply. But it is also approximately equal to the reserves of the US, which may be seen as sufficient justification for its military build-up in the area.

Western Europe

Discovery in the North Sea reached a peak in 1973, with the giant Statfjord Field. Britain exploited its share as fast as possible, partly with advantageous tax provisions, so that production peaked in 1999 and is now set to decline at about 6 per cent a year. Norway moved more cautiously, establishing a monolithic state company to take the lion's share, but its production too is now very close to peak, meaning that production in the North Sea as a whole is set to decline to approximately half its present level within ten years (see Figure I.2.1, above). It is curious that Norway, having made enormous investments in its state company, should now decide to privatize it so that foreign investors should come to own the priceless national oil and gas patrimony that is set to become infinitely more valuable as world depletion grips.

Europe's imports of oil are set to rise from the current 50 per cent to 75 per cent by 2010, and to 90 per cent by 2025, which will cause a huge drain on its balance of payments as it vies with the US and other countries for access to Middle East oil. Gas imports from Russia, North Africa, and perhaps Central Asia and the Middle East may help reduce the demand for oil until its supply comes to an abrupt, unannounced end, as explained above. Think of poor Ireland, whose demand for electricity has grown with the economic boom. It turned to gas generation, relying on a supply from Scotland, but will soon find itself very much at the end of a line from Siberia, with many energy-hungry countries in between.

Southeast Asia

India, Pakistan, Indo-China, China and Indonesia have high fertility rates, and find themselves living in a part of the world characterized by convergent plate tectonics that lack rich hydrocarbon source-rocks. China's production is expected to peak around 2003, and all the other countries are long past peak (see Figures I.2.10, I.2.11 and I.2.12). The area has been thoroughly explored, so the chance of a major pleasant surprise is remote indeed. As always, unsubstantiated claims are made for closed areas, including parts of the South China Sea, which are subject to boundary disputes. A certain, though declining, proportion of people in these countries has found out how to live sustainable lives with minimal energy demands, leaving them relatively unaffected by the decline of world oil, but Singapore, Malaysia, South Korea, Taiwan and Hong Kong are very certainly industrially advanced, energy-intensive economies, and all members of ASEAN are committed to "conventional" economic growth at the fastest rate possible.

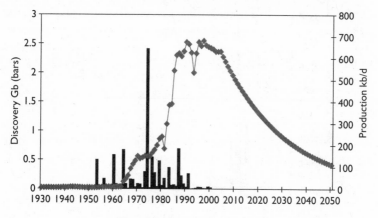

Figure I.2.10   Indian production.

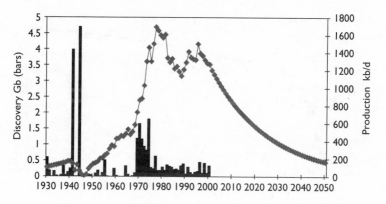

Figure I.2.11   Indonesian production.

Middle East

Lastly, we turn to review the critical role of the Middle East, which was so uniquely favored with Jurassic source-rocks and effective salt seals to hold the oil within reservoirs. These geological factors combined to make it a concentrated habitat, with most of its oil in a few super-giant fields, found long ago. Exploration has been curtailed since the expropriations, largely because the state companies, lacking the tax inducements available to Western companies, had to fund it out of national budgets, for which there were heavy competing claims. But it is worth noting that the discovery creaming curve

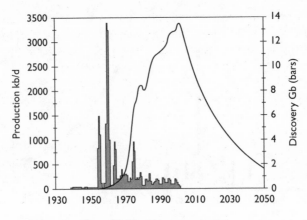

Figure I.2.12   Chinese production.

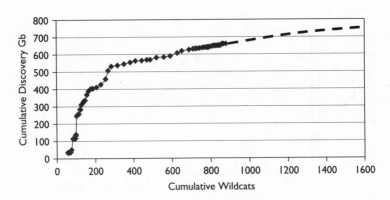

Figure I.2.13   Middle East creaming curve.

(see Figure I.2.13) has become very flat, indicating that future discovery will fall far short of past discovery.

Exactly how much has been discovered is hard to say, because the statistics are exceptionally unreliable, as already mentioned. Kuwait added 50 per cent to its reported reserves overnight in 1985, although nothing particular changed in the reservoir. Then in 1988, Abu Dhabi, Iran, Iraq and later Saudi Arabia responded with enormous increases in retaliation for Venezuela's decision to double its reported reserves by the inclusion of large amounts of long-known heavy oil.

While there are certainly skilled technicians and highly intelligent analysts in the Middle East, the management of the state companies in a highly political environment may be difficult. It is entirely possible that they remain oblivious to what their reserves truly are, possibly still relying on old reports inherited from the private companies before they were expropriated. The OPEC secretariat itself is in no position to question the information furnished to it by its member governments, much as it might be inclined to do so. The assessment is therefore to be taken with reservations. There are growing indications that the reserves are still over-stated, although this exaggeration has been recognized by officials in at least some of these countries, and reserves may consequently be revised downwards.

The degree to which the Middle East can continue to exercise its swing role is also uncertain. While the indicated depletion rates are still comparatively low, meaning that in resource terms production can be increased, there are many doubts about how much can be produced in practice. It is commonly claimed that the Middle East has much shut-in capacity, but this is doubtful. Few countries or companies have incentives to drill wells only to shut them in or choke back the production rate, except perhaps briefly, but it is only such wells, which provide spare capacity, that can be brought on at will. Infill drilling, reconfiguring wells and fine-tuning reservoir management all take work, investment and time to achieve, and the Middle East has to run ever faster to stand still as it desperately tries to offset the natural decline of its aging giant fields. It is reported that Kuwait's wells will soon be producing more water than oil, and the southern end of Ghawar (the world's largest single field) in Saudi Arabia has already gone to water. The demands on the Middle East under the best-case world scenario are illustrated in Figure I.2.14. It is far from sure if they are attainable. It is also unlikely that the position would change for the better in the event that the US invades Iraq and/or takes the Arabian fields by force. Depletion does not respond to military intervention, and pipelines are easy targets for the vanquished, even using primitive weapons.

CONCLUSION

It is not certain that Darwin got it exactly right with his view of the survival of the fittest. The experience of 500 million years of life on the planet is that species adapted to certain environmental niches

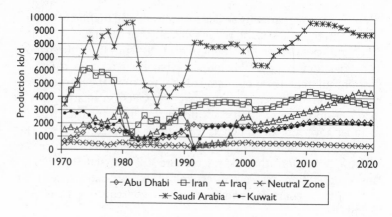

Figure I.2.14   Middle East swing production.

and proliferated, only to die out when the environment changed. The limpet, *Lingula*, which prefers a simple life attached to rocks washed by the waves, has survived unchanged since the Cambrian, but more advanced types and forms of life came and went.

Man of human appearance arrived only about 2 million years ago, and the Bronze Age that started him on the path to industrialization began only about 3,000 years ago. He did not give up his flint club because he ran out of flint, but because he found bronze made a better tool and weapon. The Iron Age followed from small and slow beginnings, but has only dramatically flourished in the last 300 years. At first this age of metals used fire-wood as fuel for smelting the metal, which in certain countries, such as Denmark and England, led to deforestation before a new fuel was found in the form of coal, lumps of which, known as sea-coal, were at first collected from beaches, before it was mined in shallow pits. Mining itself, as it penetrated below the water table, led to steam-driven machine-pumps to drain the surplus water, these pumps being later adapted to provide locomotives for transport.

The fossil-fueled heat-engine was developed into the internal combustion engine, driven at first by benzene produced from coal, before turning to petroleum refined from crude oil. This new energy form has transformed the world during the short span of a single century. Cheap and efficient transport opened the world to trade, while the manufacture of consumer goods exploded. The new energy also transformed agriculture, providing the food for a growing population that has expanded six-fold, exactly in parallel with oil production. Oil was in turn followed by gas, increasingly used for electricity generation, which brought power and light to households throughout the world, opening the door to world electronic communication, and eventually the abuse thereof through television, which helped condition the modern consumer mindset and debase human values.

This extraordinary progression was achieved in not much more than 100 years, but it was also accompanied by two world wars, together with related political repression, especially in the Soviet Union, which led to more violent deaths and suffering than the world had ever experienced before.

Now, as the twenty-first century dawns, we face the onset of the natural decline of the premier fuel that made all this possible, and we do so without sight of a substitute energy that comes close to matching the utility, convenience and low cost of oil and gas. It remains to be seen if we will be the only species in over 500 million years of recorded history to evolve backward, from complexity to simplicity. Don't hold your breath, but there is a little time left to adjust, as we have about as much oil left as we have used so far. Our challenge is to maintain demand in pace with or below the depletion rate. The first step in that direction is to determine what the depletion rate is, and to inform ourselves better about the resources with which nature has endowed us.

3

Farming and Food Production Under Regimes of Climate Change

Edward R.D. Goldsmith

Climate change is by far the most daunting problem that mankind has ever encountered. The Inter-Governmental Panel on Climate Change (IPCC) in its last assessment report predicted world average temperature rising by up to 5.8°C before the end of this century. However, it did not take into account factors such as the annihilation of tropical forests. These contain 600 billion tons of carbon, almost as much as is contained in the atmosphere, much of which is likely to be released into it in the next decades by uncontrolled logging. The Director General of the United Nations Environment Programme recently stated that only a miracle could save the world's remaining tropical forests.

Nor does the IPCC take into account the damage perpetrated on the world's soils by industrial agriculture, with its huge machines and toxic chemicals. Soil contains 1,600 billion tons of carbon, more than twice as much as in the atmosphere. Much of this will be released in the coming decades unless there is a rapid switch to sustainable – largely organic – agricultural practices. The Hadley Centre of the British Meteorological Organisation has taken these and other factors into account in its recent models, concluding that the world's average temperature will increase by up to 8.8°C this century.[1] Other climatologists are even gloomier.[2]

The IPCC and many other research agencies tell us that we can expect an increase in heat-waves, storms, floods, and the spread of human and crop diseases from the tropics into temperate areas. It forecasts a rise in sea levels of up to 88 centimeters this century, which will affect something like 30 per cent of the world's agricultural lands.[3] Furthermore, the secondary Antarctic and Greenland ice-shields are melting far more quickly than was predicted by the IPCC. This will reduce the salinity of the oceans, which must weaken, if not divert, the Gulf Stream.[4] This may lead to the freezing up of temperate Northern Europe.

It is ironic that global warming could lead to local cooling, although this has certainly happened in previous periods. Even if we stopped burning all fossil fuels tomorrow, our planet would continue to heat up for at least 150 years, while the oceans will continue to warm up for 1,000 years at least. All we can do is take measures – dramatic ones – to limit damage and slow down the warming trend, so that when our climate eventually stabilizes our planet will remain partly habitable.

Extreme weather in many parts of the world reveals that climate change is proceeding faster than predicted. Four years of drought in much of Africa have resulted in 30–40 million people facing starvation. Simultaneously, drought in the bread-baskets of the world – the American cornbelt, the Canadian plains, and the Australian wheat belt – will seriously reduce cereal exports; Australian cereal production and export capacity is already falling significantly. Floods in Germany in 2002 are expected to cost at least US$13 billion. Storms in Italy, with hailstones the size of tennis balls, destroyed crops over a wide area in 2002. Drought in the southern Italian province of Foggia has ruined olive harvests. Southern Sicily is said to be drying up.

All this is the result of no more than a 0.7°C rise in global average temperature.

EMISSIONS OF GREENHOUSE GASES FROM AGROINDUSTRIAL FOOD PRODUCTION

We must transform or completely restructure our food production system so that it helps us to combat global warming and, at the same time, feeds us. Transformation is imperative, because industrial agriculture is responsible for 25 per cent of the world's carbon dioxide emissions, 60 per cent of methane gas emissions and 80 per cent of nitrous oxide, all of which are powerful greenhouse gases.[5]

Nitrous oxide is some 200 times more potent than carbon dioxide as a greenhouse gas, though its concentration is 1,000 times lower, at 0.31ppmv (parts per million by volume) compared with 365ppmv for CO_2. Nitrogenous fertilizers are a major source of nitrous oxide. Around 70 million tons a year of nitrogenous fertilizers are currently used, and these contribute as much as 10 per cent of total nitrous oxide emissions, or about 22 million tons per year. With fertilizer applications increasing substantially, nitrous oxide emissions from agriculture could double in the next 30 years.[6]

Nitrous oxide is also generated through the action of denitrifying bacteria in the soil. When tropical rainforests are converted into pasture, nitrous oxide emissions increase by about 200 per cent, or three times. Overall, land conversion is leading to the release of around 500,000 tons a year of nitrogen in the form of nitrous oxide, or the equivalent of 50 million tons of carbon dioxide in terms of climate warming.

In the Netherlands, which has the world's most intensive farming, up to 580 kilograms per hectare of nitrogen, in the form of nitrates or ammonium salts, are applied each year. At least 10 per cent of that is immediately released to the atmosphere, either as ammonia or nitrous oxide.[7]

The growth of agriculture is also leading to increasing emissions of methane. In the last few decades, there has been a substantial increase in livestock – cattle, in particular – much of which has been made possible by the conversion of tropical forests to pasture. Cattle emit large amounts of methane (around 40 liters/day per cow) and the destruction of forests for cattle-raising is therefore leading to increased emissions of two of the most important greenhouse gases. Worldwide, the emissions of methane from livestock amount to some 70 million tons. Cattle are increasingly fed on a high-protein diet – especially when fattened in feedlots. Such cattle emit considerably more methane than grass-fed cattle. The use of nitrogen on grasslands can both decrease methane uptake and increase nitrous oxide production, increasing atmospheric concentrations of both.[8]

ENERGY INTENSITY

The most energy-intensive components of industrial agriculture are the production of nitrogen fertilizer, and the use of farm machinery and irrigation pumps. These account for more than 90 per cent of the total direct and indirect energy used in agriculture, and they are all essential to the intensive agriculture capable of supporting our endlessly growing urban-industrial civilization.

Relative to traditional agriculture, supporting smaller cities with much lower levels of industrialization, emissions of carbon from the burning of fossil fuels for intensive agriculture are often seven or eight times greater per unit of cultivation. Emissions for agricultural purposes in England and Germany, for example, are about 0.046

and 0.053 tons per hectare compared to 0.007 tons (roughly seven times lower) in non-mechanized agricultural practices and systems.[9]

This ties in with the estimate made by Pretty and Ball[10] that to produce a ton of cereals or vegetables by means of modern agriculture requires six to ten times more energy than sustainable agricultural methods.

A very large, even extreme reduction in gas emissions is immediately necessary if we accept the Hadley Centre's contention that rising temperatures within 30 years will have become sufficient to switch our main CO_2 and methane sinks (forests, oceans and soils) into *sources* of these greenhouse gases. If this occurs, we would be caught up in an unstoppable chain-reaction towards increasing temperatures. This would possibly be further intensified by massive releases of methane, presently "captured" in the form of methane hydrates in ocean sediments, of which the release is entirely dependent on temperature.

RADICALLY CHANGING AGRICULTURAL PRACTICES

We must therefore urgently develop sustainable agricultural practices that do not present these extreme risks for climate stability. The practices we need must not "mine" the soil, degrading its quality by over-intensive utilization, compacting soil surfaces by machinery, and removing organic nutrients through excessive use of fertilizers and irrigation, and so on, but on the contrary must help to maintain and improve the quality of soil resources. These "new" practices would have much in common with the practices of our ancestors, and those of communities in the remoter parts of the Third World which have succeeded in staying, to some extent at least, outside the orbit of the industrial system. They may be "uneconomic" within the context of an aberrant and necessarily short-lived industrial society, but they are the only ones designed to feed people in a *sustainable* manner. Authorities on sustainable agriculture, among them Jules Pretty and Miguel Altieri, consider it synonymous with "traditional agriculture."

If traditional agriculture is required, why are governments and international agencies so keen to prevent traditional peoples from practicing it anymore, and to force them to adopt industrial agriculture? The answer is that traditional agriculture is not compatible with the "development" model that we impose on the people of the Third World, still less with the global economy and the interests of the transnational corporations that control the "global market." In addition, given continuing population growth and urbanization, it is believed (by most UN agencies, for example) that only intensive agriculture can "win the race" against rising numbers of mouths to feed.

It is clear, however, from the following quotes from two World Bank reports,[11] that the superiority of traditional agriculture can be appreciated even by the agents of breakneck urbanization and industrialization. Thus, on the subject of development in Papua New Guinea, the World Bank admits that "a characteristic of Papua New Guinea's subsistence agriculture is its relative richness." Indeed "over much of the country nature's bounty produces enough to eat with relatively little expenditure of effort." Why then change it? The answer is clear: "Until enough subsistence farmers have their traditional lifestyles changed by the growth of new consumption wants, this labor constraint *may make it difficult to introduce new crops*"[12] – that is, "agrocommodity" cash crops, suitable for large-scale export.

Even in the World Bank's iniquitous Berg report, it is acknowledged "that smallholders are outstanding managers of their own resources – their land and capital, fertilizer and water."[13] But in the same report it is also acknowledged that the dominance of this type of agriculture or "subsistence production" "*presented obstacles to agricultural development*. The farmers had to be induced to produce for the market, adopt new crops and undertake new risks."[14]

THE END OF THE PARTY FOR INDUSTRIAL AGRICULTURE

Like it or not, "modern" fossil-energy-dependent agriculture is on the way out. It is proving ever less effective. Diminishing returns extend from fertilizer use through irrigation to the use of "transgenic" or genetically modified plants and animals. The Food and Agricultural Organisation of the United Nations (FAO) admitted in 1997 that wheat yields in both Mexico and the US had shown no increase in 13 years. In 1999, global wheat production actually fell for the second consecutive year, to about 589 million tons, down 2 per cent from 1998. Fertilizers are too expensive, and as McKenney puts it "the biological health of soils has been driven into such an impoverished state in the interests of quick, easy fertility, that productivity is now compromised, and fertilizers are less and less effective."[15]

Pesticides, too, are ever less effective. Weeds, fungi, insects and other potential pests are amazingly adaptable. Five hundred species of insects have already developed genetic resistance to pesticides, as have 150 plant diseases, 133 kinds of weeds and 70 species of fungus. The reaction today is to apply more powerful, more lethal and more expensive poisons, which in the US cost US$8 billion a year, not counting the cost of spreading them on the land.[16] Farmers are losing the battle; while the pests are surviving the onslaught of what are nothing more than chemical weapons, the farmers are not. More and more of them are leaving the land, and the situation can only get much worse.

Today we are witnessing the forced introduction of genetically modified crops by international agencies in collusion with national governments, as the result of massive lobbying from an increasingly powerful biotechnology industry. Genetically modified crops, quite contrary to what we are told, do not increase yields. They require more inputs, including more herbicides, whose use they are supposed to reduce significantly, as well as irrigation water. The science on which they are based is seriously flawed. No one knows for sure what will be the consequences of introducing, by very rudimentary techniques, a specific gene into the genome of a very different plant or creature. Surprises are in store, and could cause serious problems.[17]

TOTAL DEPENDENCE ON DIMINISHING OIL RESERVES

Another reason why industrial agriculture has had its day, even without climate change, is that it is utterly vulnerable to increases in oil prices, and even more to shortages in its availability. Recent famine and ongoing food shortage in North Korea are partly a result of Russia no longer being able to purchase North Korean export goods, with their payment in oil and petroleum products at "friendship" prices. This has – deliberately – been aggravated by the US decision to suspend shipments of oil to North Korea, as a "bargaining chip" in its attempt to force Korean compliance with the Nuclear Non-Proliferation Treaty and abandon graphite-cooled reactors that produce somewhat more plutonium than conventional nuclear reactors. North Korea can no longer afford to import the oil on which its highly mechanized, Soviet-inspired agricultural system has become dependent. Its "farmers" had forgotten how to wield a hoe or push a wheelbarrow. Many thousands of North Koreans have starved to death in the last three years.

In an industrial society, oil is required to transport essential food imports, to build and operate tractors and farm machinery, to operate and maintain farm buildings, to produce and use fertilizers and pesticides, notably from natural gas, and to process, package and transport food to supermarkets. Further energy, partly oil- and gas-based, is then needed to build and operate refrigerators or freezers for food storage, and finally to cook the food in households and transport wastes generated by its consumption. A more vulnerable, fossil-energy-dependent situation is difficult to imagine. Because of the imminence of Peak Oil, it is not just temporary oil shortages associated with jumps in the price of oil that we are destined to face, but a decline in its physical availability. Necessarily, prices will rise, whether they are expressed in a "fiat" currency or not. As this occurs oil will become increasingly expensive until it will be affordable to only a minority of corporations – US ones, in all probability, as the US oil industry is positioning itself to attempt sequestration of major residual oil reserves in the Middle East and Central Asia.

PROTECTING THE SOIL

Industrial agriculture's main contribution to carbon dioxide emissions is via the loss of soil carbon to the atmosphere,[18] caused by such practices as:

- deforestation and the drainage of peat lands and wetlands in order to make land available for agriculture and livestock;
- deep ploughing, which exposes the soil to the air and rain, and when practiced on steep slopes results in serious soil erosion;
- the use of heavy machinery that compacts the soil, reducing pore space that provides channels for air, water, plant roots and soil micro-organisms;
- the use of chemical fertilizers as a substitute for natural fertilizers, which destroys soil structure and kills soil organisms;
- the use of pesticides, some of which, as Rachel Carson[19] showed way back in 1962, do exactly the same thing;
- overgrazing, which has led everywhere to soil degradation and spectacular desertification in drier, savannah-type environments;
- large-scale monoculture, year after year, which eventually turns the soil into a lifeless dust-like substrate for crops that can only mature if dosed with increasing amounts of irrigation, artificial fertilizers and other inputs.

The most obvious method of preventing soil loss and increasing organic matter in the soil is the use of manures, compost, mulches and cover crops which can be fed back into the soil. These protect the soil from erosion, desiccation and excessive heat, and promote the decomposition and mineralization of organic matter.[20] They also reduce soil-borne diseases and increase productivity. As Jules Pretty notes, in the Niger Republic mulching with twigs and branches permits cultivation on hitherto abandoned soils,[21] "producing some 450 kilograms of cereals per hectare. In the hot Savannah area of northern Ghana, straw mulches combined with livestock manures produce double the maize and sorghum yields of the equivalent amount of nitrogen fertilizer."[22] Pretty cites other impressive examples, in Guatemala, the state of Santa Katarina, Brazil and elsewhere.

It is important that the soil should be left uncovered for as short a time as possible. An undercrop, preferably a leguminous one such as lucerne, can be sown along with a crop of cereals so that when the latter is harvested the land remains under cover, and at the same time enriched.

Conservation tillage, or better still zero tillage appears ideal as it entirely avoids ploughing. However, getting rid of weeds requires undesirable herbicides. What is needed is zero tillage *without* the use of herbicides. If the area involved is small, mulches can be used to smother the weeds. Alternative methods for killing weeds need research.

The agricultural methods required to protect our invaluable soil resources, which are essential for coping with climate change, provide many benefits. They give rise to a higher biodiversity of soil micro-organisms and micro-fauna. They are energy efficient because of their lower dependence on energy-intensive inputs. By adding biomass to the soil, they increase productivity and reduce costs, and they provide very much healthier food.

AGROFORESTRY

The FAO, in the report already referred to, tells us that the absorption of carbon by the soil is maximized by agroforestry.[23] If it were practiced worldwide, it could absorb in a ten-year period (to 2010) about 1.3Pg (Peta grams, or billion tons) of atmospheric carbon annually.[24] The IPCC, in its Third Assessment Report (2000),[25] also concludes that agroforestry yields the best results not only by increasing soil organic matter but also through increasing woody biomass.

The USDA National Agroforestry Center (2000) agrees that carbon sequestration by agroforestry is particularly effective. The Center favors short-rotation coppicing in which the wood, burnt in lieu of fossil fuel, provides a double benefit through carbon sequestration and energy substitution. The Agroforestry Center suggests that, with coppicing, soil carbon can be increased by 6.6 tons C/ha/yr (tons of carbon per hectare per year) over a 15-year rotation, and wood by 12.22 tons C/ha/yr over the rotation.[26]

Combining agriculture with forestry is a *solution multiplier*. Wind velocity over exposed soils is reduced. In summer, the temperature under trees is much lower than in open areas, and is warmer in winter. Humidity under trees is also greater, because of reduced evaporation, and improved soil structure increases water retention. The leaf litter makes excellent fertilizer when composted. Forested areas also play an enormous role in preventing floods, as rainfall is released slowly.[27] Forested areas are also a source of food and forage, as well as vegetable dyes, medicinal herbs and wood. Tree crops are valuable. The sweet chestnut has a very high food value, for instance, and was grown extensively in high altitudes in southern Europe for making flour for pasta and bread. In the tropics, perennial tree crops such as breadfruit, plantain, and jackfruit are still important.

IRRIGATION

Another essential change will involve phasing out perennial irrigation. This is one of the most energy-intensive components of industrial agriculture. Pimentel considers that when it is based on the use of water extracted from a depth greater than 30 meters, pumped irrigation requires more than three times more fossil fuel energy for corn production than rain-fed cultivation.[28] In many regions of the world, notably in India, Pakistan and the Middle East, irrigation water is often extracted from as deep as 100 meters below ground level, from aquifers that may have replenishment periods of above 1,000 years. Rice cultivation, which feeds a very large proportion of people in the tropical world, gives rise to very much more methane gas when rice fields are flooded and treated with artificial fertilizer, rather than rain-fed and grown organically. The reason is that flooding cuts off the oxygen supply to the soil, causing the organic matter it contains to decompose into methane gas.[29]

Admittedly, modern perennial irrigation is highly productive and makes three crops a year quite feasible. Indeed, about 11 per cent of the world's cropland (250 million hectares in 1994) is under perennial irrigation, and supplies as much as 40 per cent of the world's food.[30] Our dependence on perennial irrigation is largely due to the cultivation of crop varieties such as the hybrids of the Green Revolution, followed now by genetically modified varieties, always requiring greater irrigation, fertilizer and pesticides per hectare cultivated. Traditional varieties, some of which are also highly productive, require *less* water. In some areas of India farmers are returning to them, simply because of the mounting costs and diminishing returns of "miracle" GM crops. Irrigation dependence is also due to ever-increasing production of highly water-intensive agrocommodity export crops, such as sugar cane, eucalyptus, and worse still "oil-fed beef." As Reisner notes, to produce a pound of corn (maize) requires some 100 or 200 gallons of water, mostly pumped from increasing depths. But to produce a pound of beef requires up to 8,500 gallons: 20 to 80 times more water.[31]

Modern irrigated agriculture could not be less sustainable. The amount of water used for irrigation is doubling every 20 years, and at present consumes nearly 70 per cent of all the water used worldwide. This cannot go on much longer, for the twin reasons of declining hydrographic resources and fossil energy depletion, with or without climate change. Almost without exception modern irrigation, especially in tropical areas, leads to waterlogging and salinization. This sterilizes land and requires ever more to be brought under irrigation, again increasing the energy-intensity of often declining output.

In the US alone, 50–60 million acres, or 10 per cent of all cultivated land, has already been degraded by salinization, with the loss of many thousands of acres from the cultivable area in the past few decades. The depletion of groundwater resources has been just as dramatic. The massive Ogallala aquifer, once regarded as practically inexhaustible, is being depleted at the rate of 12 billion cubic meters per year. Over the years it has lost 325 billion cubic meters of water. At the world level, the annual depletion of aquifers now amounts to at least 163.6 billion cubic meters.[32] Land taken out of or lost from irrigated agriculture becomes second-rate grazing land, production and productivity fall flat, and such areas support only a fraction of their previous human populations.

More than a billion people worldwide now suffer from water shortages, and their number can be expected to increase dramatically in the coming decades. Much of the water in the world's main rivers is derived from melting glaciers; these are in full retreat as a result of global warming. By consequence the flow of many rivers will be reduced, in some cases, according to Cynthia Rosensweig, by as much as 25 per cent, and this will be aggravated by increased sea levels, which reduce river gradients from source to sea.

Also, as Bunyard notes,[33] the amount of water required for irrigation as surface temperatures rise must increase, partly because of the increased evaporation from the soil, from reservoirs and irrigation channels, but because of increased evapotranspiration from crops and vegetation. The reaction of governments and the World Trade Organization is, as usual, to transform the problem into a business opportunity! Under the General Agreement on Trade in Services, water is being privatized, and wherever this happens its price automatically rises. In the Indian state of Orissa, according to Vandana Shiva,[34] water prices have increased tenfold, and are now way beyond the means of small farmers.

ABANDONING IRRIGATION "AGRICULTURE"

The only answer is to abandon cultivation of water-intensive crops and the rearing of livestock for export. Instead we must return to *traditional* varieties of subsistence crops, most of which are rain-fed, and to *traditional* methods of irrigation, which are seasonal as opposed to perennial, and do not give rise to salinization, water-logging or the other, multiple, and increasing problems caused by modern irrigation practices.[35]

Farmers in the Malwa Plateau in the State of Madhya Pradesh in Central India, for example, are returning to un-irrigated wheat varieties, which they had abandoned under government and corporate pressure 30 years before.[36] Traditional irrigation has been practiced throughout the Indian Subcontinent, Sri Lanka, Java and elsewhere for hundreds of years. It is based on water "harvesting" and is managed by local communities in a highly democratic and sustainable way. Anil Agarwal and Sunita Narain tell us that during the drought of 1987, remote villages which had not "benefited" from government water schemes continued to have water because their traditional water harvesting systems had remained intact. In the "developed villages," on the other hand, wells had either no water or no electricity for the pumps. Agarwal and Narain also tell us how Jodhpur, the famous desert city, once had an astounding water-harvesting system with nearly 200 water sources – about 50 tanks, 50 step wells and 70 wells. In addition, people used to collect the rainwater from rooftops.[37] The surrounding catchment areas were once covered with thick forest abounding in wild animals. Today the forest has gone and the tanks – beautiful structures as they were – are used as refuse dumps. When modernization brought a piped water supply, Agarwal and Narain note, "they came to neglect their traditional systems and to depend on the government"[38] – yet another policy that must be reversed. The tanks must soon be restored, and communities must organize themselves to manage them as they once did. There is *no* alternative.

LOCAL FOOD

Farmers must produce food for local consumption, instead of for export which they are forced to do by the IMF and the World Trade Organization. Transport alone accounts for one eighth of world oil consumption,[39] and the transport of food accounts for a considerable slice of this. Over 83 billion ton/kilometers of food products and animal feeds are brought into the UK annually by sea, air and road, and this requires 1.6 billion liters of fuel, producing annual emissions of 4.1 million tons of CO_2.[40]

Air transport is the most energy-intensive form of transport. To give an idea, about 125 calories of fossil energy (aviation fuel) are needed to transport one calorie of lettuce across the Atlantic.[41] Unfortunately, more and more food is being transported by air; indeed, since 1980 imports of fruit and vegetables by air-freight into the UK have increased by nearly four times. The Royal Commission on Environmental Pollution has estimated that, on current trends, the contribution of air transport to man-made global warming is expected to increase by five times between 1992 and 2050.[42] Scandalously, the UK government promotes this by exempting airlines from both fuel tax and value added tax. Airlines pay up to four times less for fuel than anyone else.[43]

According to a study carried out in 2001, greenhouse gas emissions associated with the transport of food from local farms to farmers' markets are 650 times lower than the average for the central purchase and distribution of food products to supermarkets. In addition, to produce food locally, as the same report notes, "would be a major driver in rural regeneration as farm incomes would increase substantially." There would also be more cooperation among local people, and communities would be revitalized.[44]

Local production and distribution of food are necessary even without climate change, for it is only by producing food locally that the poor, particularly in the Third World, can have access to it. One of the main causes of malnutrition and hunger in poor countries is the shortage of land. Anything between 50 per cent and 80 per cent of the agricultural land of Third World countries is now geared to agrocommodity exports. Local people are reduced to growing their personal food requirements on marginal, rocky outcrops or steep slopes. Urban Jonsson, the UNICEF country representative in Tanzania, tells us that, "when the world economy and Tanzania's State economy are doing well, the villagers sell much of their maize and other staple foods. But when the State economy is in a bad way ... prices for food drop and give the farmer less incentive to sell. Thus the villagers do the only thing possible – they keep the food and eat it themselves." They also use land which they previously used for cash crops to grow food for their own consumption. In other words, it is only when they cannot export their food that they eat properly.[45]

RELATIVE SELF SUFFICIENCY

To produce food locally requires increasing self-sufficiency, from the ground up, and from the village, through regional levels, up to that of the state. It also means storing food at each of these levels in order to face possible food emergencies, which scandalously enough is illegal today, as the WTO considers that the money required is better spent on paying debts to Western banks. The way that international agencies define "self-sufficiency" has *nothing* to do with the real meaning of the term, these agencies affirming, for example, that a country which produces no food at all can be regarded as "self-sufficient" as long as it can pay for its imports.

What we call food self-sufficiency they decry as "food autarky," for them the greatest crime any country can possibly commit. The reason is that if food autarky were adopted worldwide there would be almost no international agrocommodity trade, the global economy would immediately shrink, and many transnational corporations would disappear as international trade shrank to that in food surpluses, manufactured goods, and services. Without the prop of cheap biological resources, imported under flagrantly unfair terms of trade from the poorer countries, the vast majority of OECD countries would themselves be forced to restructure their economies, societies and culture. But it is imperative that, at a world level, we shift to something approaching food autarky, or rather self-sufficiency. This will be essential – though not in the extreme sense of the term, as some trade will always be beneficial; but it is largely *surpluses* that should be traded.

SMALL FARMS

Farms that cater to their local area and are largely self-sufficient must necessarily be small. Big farms, to survive, must cater to the world market as they increasingly do, or become uneconomic and cease operations. Moreover, to maximize "economic efficiency" as it is currently conceived, farmers have no choice but to use heavy machinery, fertilizer, pesticides and irrigation water, eliminate hedgerows and tree cover, and produce one or a few cash crops over vast "monocrop" expanses of land, year in and year out. This is exactly what we need to avoid – even without climate change and fossil fuel depletion. We also need small farms because they are much more *productive* than big ones. Even the FAO, which has spearheaded the shift towards industrial agriculture worldwide,[46] now admits this. Thus an FAO report makes clear that the farms with the highest productivity in Syria, for instance, were found to be about 0.5 hectares in size, while in Mexico 3 hectares, in Peru 6 hectares, in India less than 1 hectare and in Nepal a little less than 2 hectares. In each case output was found to fall as soon as the size of the farm increased beyond these levels.[47]

The most productive form of food production is undoubtedly horticulture. In the UK, according to Kenneth Mellanby,[48] an English vegetable garden can produce as much as 8 tons per acre each year. Significantly, during World War II, 40 per cent of Britain's food, including vegetables, was derived from just over 300,000 acres of vegetable gardens and allotments (albeit for a population some 20 million smaller). Unfortunately most of these allotments were situated close to urban centers and have since been "developed." Clearly they should urgently be restored, and extended.

One reason why productivity is so high in a small farm or garden is that the most important input, as Dr. Schumacher always put it, is TLC – "tender loving care." This traditional but effective "ingredient" is far more likely to be applied by small farmers, who totally depend on their land for their livelihood, than large-scale commercial farmers, who are only in it for the money. With climate change, of course, ever more TLC will be required.

DIVERSITY AND VARIETIES OF CROPS

A localized, largely self-sufficient farming system, made up largely of small farms, is necessarily based on cultivating a *wide variety* of different crops, and even different varieties of these crops, as traditional farmers have always done. In addition, as Peter Rossett notes, small traditional farmers often intercrop the spaces between rows, enabling weed numbers to be reduced, and the simultaneous rotation of livestock and cropping to be facilitated.[49] Jose Lutzenberger, who was once Minister of the Environment in Brazil,[50] tells us that the Italian and German peasantry that established itself in South Brazil cultivated many different crops, including sweet potatoes, Irish potatoes, sugar cane, cereals, other vegetables, grapes and all kinds of fruit, while also producing silage for their cattle, as well as rearing chickens, pigs and cows. The total production of these small farms amounted to at least *15 tons* of food per hectare, incomparably more than is produced by modern soybean monocultures in the same area, all of which depend entirely on chemical inputs. What is more, there is a strong synergistic relationship between the different crops cultivated by these traditional farmers.

Thus, in a well-planned inter-cropping system, early established plants tend to produce the appropriate microclimate for other plants. Plants also complement each other in terms of nutrient cycling; deep-rooted plants can act as "nutrient pumps," bringing up minerals from the subsoil. Minerals released by the decomposition of annuals are taken up by perennials. The high nutrient demands of some plants are compensated for by the addition of organic matter to the soil by others. Thus cereals benefit from being grown in conjunction with legumes, which have deeper roots, permitting a better use of nutrients and soil moisture, as well as possessing root nodules which host bacteria specialized in fixing nitrogen. Crop diversity thereby plays a significant role in the metabolism of a traditional agricultural ecosystem, and contributes to its productivity. However, if traditional small farmers plant such a wide diversity of crops, it is not primarily to maximize yields but to reduce vulnerability to discontinuities such as droughts, floods and plant epidemics.

As James Scott, an authority on peasant agriculture, writes, "the local tradition of seed varieties, planting techniques and timing was designed over centuries of trial and error to produce the most stable and reliable yield possible under the circumstances." Typically, the peasant seeks to avoid the failure "that will ruin him rather than attempting a big but risky killing,"[51] and this he largely achieves by cultivating a carefully chosen diversity of crops and crop varieties, whose exact composition he is well capable of adapting whenever necessary to changing environmental requirements.[52]

A de-industrialized world in which people live in small towns and villages, and produce much of their own food and artifacts locally, would be largely unaffected by oil shortage. It would also be incomparably healthier. There would be far less poverty and hunger, and fewer wars, as the majority that have been fought in the last 100 years were partly or wholly triggered by competition for access to the markets and resources that *only* globalized industrial society requires. Nor of course would its economic activities destabilize the atmosphere.

ELIMINATING ARTIFICIAL FERTILIZER: A SOLUTION MULTIPLIER

Every measure that brings agriculture closer to traditional methods is a solution multiplier. Consider the host of problems created by artificial fertilizers. By replacing them with natural fertilizers we would solve a number of serious problems:

1. Artificial fertilizers reduce the capacity of the soil to absorb carbon dioxide by disrupting soil ecosystems and according to P.A. Steudler this also applies to the absorption of methane gases.[53]
2. When washed into rivers and estuaries, they stimulate deoxygenating algal bloom (eutrophication).[54]
3. The algae often form huge masses which emit dimethyl-sulfide, which oxidizes in air to sulfur dioxide, the principal source of acid rain.[55]
4. Fertilizers are the largest source of pollution of ground and drinking water.
5. Fertilizer use can raise nitrate levels in food to unhealthy levels.[56]
6. Nitrates are transformed by bacteria into nitrites, which bind to hemoglobin and reduce the ability of blood to transport oxygen,

often giving rise to methemoglobinaemia, a blood disorder of young children.[57]

7. Nitrates when, combining with amines, in the gut can be transformed into carcinogenic nitrosamines.[58]

8. Organic foods and food produced without fertilizers contain an increased range and quantity of phytonutrients, and have higher vitamin C and mineral content. The plants can better withstand pests and diseases. Trials have shown improved growth, reproductive health and recovery from illness of animals fed organic feed.[59]

9. Studies at the Obervil Institute in Switzerland have shown that wine grape yields can only be increased by maximizing nitrogen applications at the cost of reducing their sugar content, which prevents them from ripening properly. Studies at the Biodynamic Research Station in Sweden found that potato yields could be increased by 15 per cent if enough fertilizer was applied, but this drastically increased post-harvest losses during storage.

In Sri Lanka traditional farmers note that they have no difficulty in storing traditional strains of rice for three to four years, while the hybrid varieties using artificial fertilizer become moldy in three months.[60] The probable reason is that higher nitrate applications create a problem for the plant by increasing the osmotic pressure on cells, and to accommodate this the plant must extract more water. Thus the yield of a compost-grown plant was found to be 24 per cent lower, but its dry matter was 23 per cent higher. Fertilizers did *not* increase the dry weight, but simply added more water to the crop. Consequently, artificial fertilizer produces watery crops vulnerable to fungal infestations and post-harvest losses. A higher use of poisonous pesticides is thus necessitated during storage.

The much-vaunted benefits of artificial fertilizers are illusory. This is not altogether surprising, as they were not developed in the first place for the purpose of providing people with cheap, plentiful and healthy food. They were designed as explosives (TNT), forming a complement to the pesticides and herbicides of the "scientific" agroindustry that were initially developed as chemical weapons. These include, for example TEPP (tetra-ethyl pyrophosphate), and the herbicides 2,4-D and 2,4,5-T of which the US sprayed over 900 million liters in Vietnam as a "resource denial" weapon, resulting in at least 75,000 deaths and continuing birth defects some 28 years after the war's end.

FROM GREEN REVOLUTION TO "GENE REVOLUTION"

The Green Revolution imposed by America on the Third World was part of a campaign to sell more fertilizer and keep the armaments industry afloat after World War II. Its high yielding varieties (HYVs) should in fact be called "high response varieties" (HRVs) – varieties designed to be highly responsive to fertilizers, herbicides and pesticides. Many traditional varieties provide equally high yields *without* fertilizers, herbicides or pesticides.

Similarly, the "Gene Revolution" is a means of creating dependence on, and selling more pesticides and herbicides. Some 60 per cent of genetically modified varieties marketed so far are designed for resistance to herbicides such as Monsanto's "Round-Up," rather than to the diseases themselves, drastically increasing the markets for these poisons, which can now be used on crops (soy, beet, and so on) which would not previously have tolerated them. It can be argued that the overriding goal of the "biotech industry" is to control the world's entire food production process. How better to do this than by controlling the seeds on which the whole process must depend, as argued by José Bové?

Fertilizers are not just promoted on their own but as part of a *package* that includes hybrid and genetically modified patented seeds, pesticides, heavy machinery, and water derived from perennial irrigation. All of these create serious problems, are totally dependent on fossil fuels, and in combination are completely unsustainable. This conclusion is supported by the observed fact that diminishing returns on fertilizer use, and from other "hi-tech inputs," are now being experienced just about everywhere. With the coming world water and oil shortages, the use of fertilizers, like all off-farm inputs to modern agriculture, can only become ever less attractive. Their use must seriously decline, while the traditional methods discussed above are rapidly introduced or reintroduced.

4

The Laws of Energy

Jacob Lund Fisker

Just as money is an instrument used to change and describe rates of change in the production, use and ownership of goods and services by humans, energy is an instrument used by scientists to describe how states of matter change in nature. While money is subject to political and economic manipulation, the laws governing the flow of energy cannot be bent or broken. The increase in human wealth and well-being during the past few centuries is often attributed to such things as state initiatives, governmental systems and economic policies, but the real and underlying cause has been a massive increase in energy consumption. Before the Industrial Revolution human societies were necessarily based on "steady-state economics," due to the limited flow of their solar-derived energy supplies. With the advent of fossil energy supplies economic growth was enabled, and "growth economies" remain dependent on a framework of cheap and increasing supplies of energy.

The downstream money value of "stock" energy supplies – that is, coal, oil, gas and uranium – is determined by market availability; fossil fuels are cheap because relatively small investments of energy in machinery and labor bring massive returns of easily utilized energy, also giving economic markets and political leaders much freedom in setting price levels for those stock resources. Initial return on investment with stock resources is huge compared to the laborious exploitation of "flow" sources of energy, because discovering and extracting fossil fuels requires little effort when resources are abundant, before their depletion. It is this cheap "surplus energy" that has enabled classical industrial, urban, and economic development. As the world's energy stocks are depleting, making what is left harder to extract, it becomes less worthwhile spending money, human time and effort searching for and extracting ever-smaller deposits of fossil fuels, as is shown by ever-decreasing oil and gas exploration efforts, even with rising prices for oil and gas. After Peak Oil, however, yet more effort and more resources will have to be put into building new technology to exploit the lower-yielding, renewable, or "flow" resources, leaving *less* "surplus energy" for the rest of the economy.

The availability of money determines the distribution of goods. With less "surplus energy," there will be fewer products made and distributed for money to buy. Only two economic results are possible – money inflation or economic slump; that is, contraction of the economy. This means that the politics and economics of a world with declining energy availability will operate in an entirely different way – money will either be chasing fewer and fewer goods and resources; or production of goods and resources will decline, perhaps precipitately; or both.

The classical "laws" of economics underpinning our systems of production and exchange are essentially based on conceptions developed in the period from about 1750 to 1860 in which science and engineering presupposed that all energy sources are mutually substitutable and convertible. This fundamental misconception concerning energy was later corrected, notably by the body of knowledge called *thermodynamics*. Scientific investigation and theory became able to demonstrate physical limitations on one kind of energy source being substituted for another, yet economic and popular conceptions of energy have unfortunately never caught up.

WHAT IS ENERGY?

A first glance reveals many different forces in nature. Think of the tensile force of a piece of string, the force of a spring, the intermolecular forces holding a water drop together, the forces acting between atoms in a molecule, or the gravitational force attracting stars together to form galaxies. Other forces act between the protons and neutrons binding them together in atomic nuclei, or cause radioactivity. These are just a few examples. A closer examination shows that all these examples can be attributed to just four forces: namely the gravitational force, the electromagnetic force, the strong nuclear force, and the weak nuclear force.

Merely describing how various forces act on a piece of matter will not explain its motion. The theory connecting forces acting on particles of matter with their resulting motion is called mechanics. Isaac Newton was the first to propose a scientific theory of mechanics. Other and more refined theories of mechanics, quantum mechanics and general relativity are necessary to explain gaps in Newton's theory of mechanics – such as the observed discrepancies in extreme cases concerning very small objects, very dense objects, and very fast-moving objects – but for ordinary low-speed objects Newton's laws are as valid now as they were more than 300 years ago.

It happens frequently that the force acting on an object depends on *how* the object is placed in relation to other objects. This makes a straightforward application of Newton's laws very difficult or impossible. Mechanics therefore introduced the abstract concepts of *work* and *energy*, which have well-defined and exact meanings in physics, unlike their use in everyday language.

According to the laws of mechanics, the amount of work done by any force is proportional to the force exerted and the distance through which the force acts. That is, the work done depends on the force exerted and how far the considered object is moved. Since many forces act on most objects, the *net force* is the sum of all those forces. An object changes its velocity when a net force is applied to it. As the force does work on the object, the energy of the object must change. The energy which the object receives is called kinetic energy (from Greek *kinein*, "to move") and it is equal to the work done on the object. This energy can be thought of as stored for later release, when opposing forces slow the object down.

In many situations we can calculate or measure the work done by certain types of forces, and give each kind of "energy" or "work done" a special name. Examples include nuclear energy, which derives from the strong nuclear force between protons and neutrons, chemical energy deriving from the electromagnetic force between electrons of the atoms forming chemical molecules, and sound, vibrational or wave energy, from the collective motion of matter due also to electromagnetic force.

Claiming that all forms of energy can be converted or "translated" between each other is a sweeping generalization, and has several pitfalls. It is not always possible to convert energy from one kind to another. It also must constantly be kept in mind that the concept of energy is just *an abstract tool* enabling mathematical shortcuts for calculations on energy systems which themselves can be entirely theoretical and impossible in reality. Through calculation we can however estimate the energies involved, and find out how much work could be done *in principle*; but this tells us nothing as to the feasibility and practicality of doing the work we calculated as available. For that, we need the laws of thermodynamics, which tell us *which kinds* of energy transformations are possible.

SYSTEMS OF ENERGY

Energy calculations require a conceptual model in which there is an imaginary boundary across which energy and matter flow; the boundary demarcates an *energy system* from its surroundings.

A system may be *open* (energy and matter may leave or enter the system), *closed* (only energy may leave or enter the system), or *isolated* (energy and matter may neither leave nor enter the system). Human society may be regarded as part of an open system. The flow resource of radiation from the sun strikes the Earth's surface and provides energy, which plants and animals convert into food. Solar energy also drives climate and weather patterns from which wind-power, and rain for hydroelectric installations, can be exploited. Mining coal and extracting oil and gas yields additional energy resources, ultimately depending on flow energy resources of the geological past. After use, low-grade energy escapes the Earth in the form of infrared heat radiation.

Mineral deposits are only "renewable" on tectonic time-scales, and fossil fuels require millions of years for regeneration. It is therefore better to consider human society as a closed system (we note that no civilization has lasted more than a few thousand years). Concerning recycling of waste minerals and energy, all such resources would eventually be exhausted without constant, fresh or new inputs to the human system. Without access to past fossil resources, our human economic system would once again depend solely on the radiative energy flux of the sun, which has enough nuclear "fuel" in its center to burn for another 4–6 billion years.

THE LAWS OF THERMODYNAMICS

See Fig I.4.1 at end of this chapter for a summary of the four Laws of Thermodynamics. An energy system is described in terms of its *state*. A description of the state may be incredibly detailed, including a microscopic description of the specific type, position, and velocity or energy of all the atomic particles in the system – in which case it is called a *microstate*.

The information content of a microstate is huge, but it can be reduced to a few macroscopic observables – such as *pressure*, *density*, and *temperature* – in which case it is called a *macrostate*. This idea is similar to the statistical idea of reducing a large sample to a few parameters, such as the mean, deviation or variation, and skewedness. In an energy system many different microstates may have the same macroscopic characteristics, and one macrostate can *represent* a huge number of microstates. The way in which a system changes from one macrostate to another is called a *process*. In some situations there may be several different *processes* by which a system can change from its initial macrostate to its final macrostate. It may also be that the *final* macrostate is identical to the *initial* macrostate of the system, and in this case the process is called *cyclic*. Cyclic processes are especially prevalent in combustion and other engine designs which use thermal energy to do work. The most famous cycles are the Otto (gasoline engine) cycle, Diesel engine cycle, Stirling cycle, Rankine or steam engine cycle, and the Carnot-cycle. It is the challenge of engine designers to maximize engine efficiency, also known as the *duty cycle*.

When a hot object and a cold object are in contact, thermal energy or *heat* is transferred from the former to the latter until they reach *thermal equilibrium*, at which point they have the same *temperature*.[1] Having realized in the late eighteenth century that heat was a form of energy, scientists of the nineteenth century soon investigated how to convert heat into work, or motion, with the greatest efficiency. This was the birth of what we call *thermodynamics*.

The first and second laws of thermodynamics comprise some of the most fundamental laws of nature, and are applied in a wide variety of physics disciplines. Almost every scientific test or experiment in physics implicitly tests the laws of thermodynamics and – so far – they have never been shown to be false.

ENERGY AND THE FIRST LAW OF THERMODYNAMICS

Explanation of this law demonstrates, using the example of an internal combustion engine, how this law *severely restricts* the possible final states of any energy transformation process. Only those final states which have *the same* total energy as the initial state are permitted by this law, thus *it is impossible to create energy out of nothing*.

A heat-driven engine is an isolated system, where gas in a combustion chamber or engine cylinder is confined by a piston. If heat is transferred to the gas, the gas will expand, moving the piston, and work can be done by the engine system on its surrounding mechanical linkages. The specific process – the *method* of heating the gas and executing and causing the power stroke to take place, of doing the work – depends on the technology and design of the engine. In an internal combustion engine like that of a car, the heat comes by igniting a compressed fuel–air mixture directly inside the cylinder. In a steam engine, conversely, the heat source is external, as the steam is generated in a boiler and led into the cylinder by a slide valve.

The amount of work done *by* the system which can be extracted from the heat transferred *to* the system varies with the specific process. The difference between the heat transferred and the work done depends only on the initial and final macrostates. Examining this difference needs the introduction of a new quantity, called the *internal energy* of the system. If not all heat is converted into work, the energy difference left in the system will increase its internal energy.

The internal energy of a gas is hidden in the random motion, rotation, and vibration of the constituent molecules as they speed around at average velocities of around 600m/s (about two times the speed of sound) at room temperature, violently colliding with the walls of the container and also with each other. The cumulative effects of the huge rate of collisions is felt macroscopically as pressure, whereas the average speed of the particles is related to the temperature. Thus heating a gas increases its internal energy and temperature, which increases particle speeds, raising both the frequency and force of collisions, corresponding to an increase of pressure. Gas pressure exerts force on the piston and causes it to move, but, as the piston moves, particles hitting the piston will transfer part of their kinetic energy into the motion of the piston, thereby cooling down the gas. Eventually the gas will cool down to a level where, despite having residual internal energy, it *cannot* push the piston any further because of higher opposing forces on the piston from its surrounding mechanical linkages – bearings, joints, and the connecting rod.

Although this description may sound quite complicated at the microscopic level, a cylinder with a piston is basically just a method of converting the somewhat useless and randomly directed motion of hot gas particles into the useful, directed motion of the piston, which can easily be converted into work.

The first law of thermodynamics states that **in an isolated system the increase of internal energy is *independent* of the process and equals the difference between the heat transferred to the system and the work done on the surroundings by the system**. In other words **heat is also a type of energy, and total energy of the system + surroundings is constant.** Thus the first law severely restricts the possible final states of *any* transformation of energy. Only those final states, which have the same total energy as the initial state, are permitted by this law. It also follows from this law that *it is impossible to create energy out of nothing.*

Even then only some of the different final states allowed by the first law are found in nature, namely because of the problem of converting the undirected random motion of the gas particles into the directed motion of the piston. The second law explains why.

ENTROPY AND THE SECOND LAW OF THERMODYNAMICS

Entropy is a measure of the random motion in a system and therefore a measure of the *limit* on the amount of work (that is, directed motion) which can be extracted from an energy system. The second law of thermodynamics states **that entropy *always increases*, and therefore more and more energy becomes *unavailable* for work.**

Consider a gas inside a container like the cylinder described above. Each time any two particles collide they exchange energy, but due to the nature of the inter-particle force (the electromagnetic force) it can be shown the *total energy* of the two colliding particles remains constant as long as no chemical reactions occur. Therefore the total internal energy of *all* the particles remains constant, as required by the first law.

In dealing with very small entities it is necessary to use quantum mechanics. Here, the Schrödinger equation is the modern equivalent of Newton's laws. Solving the Schrödinger equation for a gas with a fixed total energy results in an enormous amount of solutions, each one defining a possible and energetically allowed microstate. These microstates are subsequently grouped into different macrostates according to the pressure, density and temperature they correspond to. To pick some extreme examples of valid solutions, particles could be moving in unison back and forth between just the two cylinder walls, exerting a specific calculable pressure only on these walls through their bombardment; conversely they could be located close to a corner of the cylinder, exerting little measurable pressure. Most likely, however, particles will move around *randomly*, exerting about the same pressure on all internal surfaces. This macrostate is much more likely because it allows so many more ways for particles to be arranged.

It can be shown that, for a large number of particles,[2] *one* of the possible macrostates corresponds to an overwhelming number of microstates relative to any other macrostate and its constituent microstates. The number of microstates corresponding to this macrostate is directly related to its *entropy*, which is set by Boltzmann's law: $S = k_B \log W$, where S is the entropy, W is the number of microstates representing the "statistically favored" macrostate, and k_B is Boltzmann's constant. It is a consequence of the H-theory of statistical physics[3] that a system in thermodynamic equilibrium is equally likely to be found in *any* of the energetically allowed microstates. Therefore the most likely macrostate is also the macrostate with the *maximum entropy*, and the final state of any energy process will be the state of maximum entropy.

System entropy increases by *any process* which increases the number of microstates, including:

- a rise in temperature (heat or friction);
- an increase in volume (expansion);
- a phase transition (melting, evaporation, or sublimation);
- a chemical reaction increasing the number of molecules present.

The second law generalizes this by stating that **for any process in an isolated system the number of energetically allowed microstates never decreases,** or, **the total entropy of the system + surroundings never decreases.**

The second law complements the first law, which determines which final states are theoretically possible, by determining *which* final states can exist in practice. So while the energy conservation principle of the first law would allow cold water to heat up spontaneously while simultaneously forming ice cubes, the second law forbids this process because it would *decrease* entropy. However, neither law says anything about how fast a process will run or even if it will occur. These two laws, in combination, solely limit the processes which are *capable* of occurring.

Entropy can decrease in an *open* system, but only at the expense of a larger increase in entropy taking place in its surroundings. Photosynthesis is an example of that, increasing the order of a plant (system) using the energy of the sun (surroundings); the sun itself increases its entropy through nuclear reactions in its core, supplying the energy for the plant on Earth. Similarly the entropy inside a refrigerator (system) is decreased by the entropy increase in the fuels (surroundings) which were burned to supply the electrical energy to run the refrigerator. Entropy is also a measure of the energy which is unable to do work, i.e. *waste heat*. The significance of the second law is therefore that *any* isolated system will finally run out of energy for doing work because the system ultimately cannot recreate the fuels it uses. Therefore a sustainable system cannot be isolated. Within such a system any work done, and its own system processes, can only occur at the rate at which energy is supplied from its surroundings, and the rate at which system entropy is transferred to those surroundings.

WORK, FUEL, AND THE HEAT ENGINE

Work is the equivalent of a directed force acting through a distance, and practically all technology uses directed work in some form. Work is applied to hoist, lift, drill, saw, hammer, punch, bend, and so on – to rearrange the forms of matter. Work is easily converted into kinetic energy, and can therefore be used to move matter. This principle is applied in cars, trucks, trains, boats, and planes. Through a dynamo or other electro-mechanical devices work can be transformed into electricity which can provide light, run radios, computers, and other electronic equipment. Work can also be used to increase the pressure in a gas in order to heat it up and drive chemical processes.

There are various ways to do work. The most "primitive" method is to eat food, which contains energy stored in the chemical bonds of fat, protein, and carbohydrates, and then use muscle power. However, human physiology restricts the amount of work which can be done to about a maximum of 180 watts, making it impossible do more work just by eating more food.

Prior to industrialization various attempts were made to do more work – primarily by using flow-dependent resources such as slaves, draft animals, and water wheels. However, with the extraction of coal and the invention of the steam engine it became possible to use stock resources, and to cease depending on short-term stock resources like mature forest wood. Later, oil and gas stock resources were also added to the mix. The steam engine is just one specific type of the greater class of machines collectively known as *heat engines*, which today account for about 80 per cent of the world's electricity generation and about 90 per cent of all transportation.

The heat utilized by a heat engine usually comes from one of mankind's oldest inventions: *fire*. Fire is a chemical process requiring an initial supply of energy – the friction of a bow-drill, the discharge of a spark plug, or the heat of a match. This energy breaks the chemical bonds of a fuel molecule (such as methane gas or octane molecules in gasoline) and the surrounding oxygen, which snap together and form new and more tightly bound chemical configurations (primarily carbon dioxide and water), which speed off in random directions with the liberated energy. As the resulting mobile particles hit other fuel molecules or oxygen molecules they break these apart, repeating the process and sustaining combustion.

Another source of internal energy is nuclear fission, which is based on similar principles. As this matter is already "warm" (radioactive) it self-ignites if its density exceeds a critical value. Emitted neutrons smash into other nuclei breaking them apart, releasing yet more energetic neutrons in a chain process, moderated in nuclear reactors by some absorbing agent such as graphite or water.

Heat engines generally cycle through four processes:

Compression stage: Here a gas is compressed either by a piston moving back (gasoline, diesel engine) or by a compressor (gas turbine, jet engine). This process requires work to be applied.

Heating stage: In an internal combustion engine or a gas turbine the gas is mixed with fuel and ignited. In a steam engine or turbine the gas is heated by an external source such as coal, fuel oils, or nuclear isotopes. This increases the pressure further as the combustion provides internal energy corresponding to Q_{heat} – the energy released by the combustion.

Power stroke: The gas expands and releases energy by doing work, W, by driving a piston forward or rotating the blades of a turbine.

Cooling stage: Finally the gas must be cooled or exhausted from the cylinder, and substituted with cold gas from the surroundings in order to return the engine to its initial state. This means that the energy Q_{waste} leaving the system[4] is *unavailable* to do work.

After completing a cycle the engine (system) is returned to its initial state. Therefore the net change in internal energy is zero. In compliance with the first law the work done is equal to the heat generated by combustion minus the waste heat which was exhausted ($W = Q_{heat} - Q_{waste}$). The engine efficiency ($e = W/Q_{heat}$) defines the ratio between the work that was done to the fuel that was paid for and burned. In co-generating power plants a part of the waste heat is used for residential heating, which makes the payoff somewhat better.

While the first heat engines were very inefficient, inventors quickly improved on the design, hoping that with continued progress it might be possible one day to approach 100 per cent efficiency, converting *all* the internal energy of a fuel or heat source into useful work. This would theoretically permit, for example, the tremendous reservoir of heat in the world's oceans to provide a practically free source of energy, and direct solar energy to drive all kinds of heat engines. Unfortunately it turned out that the second law (which was unknown until the 1860s) places a severe restriction on these dreams. In the nineteenth century Carnot devised the theoretically most efficient engine: a conceptual engine with a cycle of four *reversible* processes, operating between two heat reservoirs having fixed temperatures. Many conditions must be fulfilled for a reversible process. The system must be in complete thermodynamic equilibrium with its surroundings during the process. Additionally, any increase in entropy due to viscosity, friction, inelasticity, electric resistance or magnetic hysteresis will make the process irreversible.

Kelvin and Clausius later demonstrated that the efficiency of such theoretical engines is limited by the second law and only depends on the temperature of the exhaust environment and temperature of the input or generated hot gas. This conclusion followed from fundamental laws of physics. Despite any technological process, therefore, no heat engine can ever exceed this limit. Another factor limiting the achievable efficiency of heat engines is that any reversible process must run infinitely slowly. Faster processes break the equilibrium condition, with friction or other effects making the process irreversible. As the *rate* of the cycles goes towards zero, so does the power of the engine, giving a trade-off between the efficiency and power of any heat engine.

Other types of engines than heat engines exist, depending on flow resources like water, wind, and radiation. For example, hydroelectric plants convert the gravitational energy of water into work and electricity, wind turbines exploit the kinetic energy of wind, and photovoltaic cells directly convert solar radiation into electricity. Fuel cells also convert hydrogen directly into electricity – however, no stock resources of hydrogen exist, so hydrogen must first be synthesized using energy resources from the Earth's stock or flow resources. In addition, no energy technology has any practical interest unless it provides more energy than it uses on an all-round system basis, taking account of energy requirements for their construction, maintenance and replacement.

CONCLUSIONS

In 2000 about 86 per cent of the world's energy consumption was based on depleting stock (fossil fuel) resources, 7 per cent on the draw-down of stock uranium, and only 7 per cent on flow resources, primarily hydroelectric.[5]

The discovery and subsequent exploitation of fossil fuels is the basis of virtually all food production, industries, cities and modern society. Fossil fuel-based fertilizers and pesticides make it possible to maintain the world's agricultural output, and crude oil may be formed into a host of useful materials and products, including plastics, medicines and construction materials. Despite these unique qualities and ultimately limited stock, crude oil and natural gas molecules are mostly burned to supply cheap energy. About 10 per cent of world production of oil and gas, by weight, is used for petrochemicals and plastics, much of which is simply thrown away after first use.

The energetic "cost–benefit" of fossil fuels is very large compared to any other energy resource. Currently very little effort (work) is required to extract the fuels from the ground compared to the energy content of the fuel. This contrasts with other sources of energy (photovoltaic, wind power, hydroelectric power, wood, and uranium). Unlike flow resources, where unused energy has to be stored at a loss, fossil fuels store easily. In addition, heat engines running on fossil fuels provide the greatest and most scalable power-to-weight ratio of any fuel–engine combination. This makes fossil fuels hard to beat in terms of versatility and price. Without continued access to this cheap resource it would most likely not be possible to maintain the world as we know it.

THE FOUR LAWS OF THERMODYNAMICS

0: Heat energy always flows from a hot object to cold object.
1: The total energy of an isolated system is conserved.
2: The total entropy of an isolated system never decreases.
3: The entropy of perfectly ordered matter at zero absolute temperature is zero.

APPENDIX: INSIDE AN INTERNAL COMBUSTION ENGINE

In a car the force that turns the wheels is mechanically linked back to the engine and connected to its pistons. One might imagine running the car by pushing the pistons by hand, but even for an average car that would require the strength of several hundred hard-working people. Instead, the hot gases from burned gasoline push the piston. This is conceptually illustrated below.

The molecules in air, or a mixture of gases after explosion or burning, are in fact about a hundredth of a millionth of a meter apart on average. They can be imagined as small

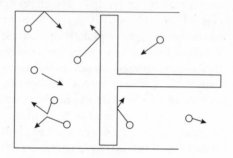

Figure I.4.1

balls not much bigger than the atoms making up the molecules, about a tenth of a billionth of a meter in size. These balls continuously collide with each other and the walls, like a frictionless game of three-dimensional billiards.

A molecule inside the cylinder will knock the piston outwards whenever it hits the piston. This increases the speed or energy of the piston while decreasing the energy of the molecule by the exact same amount. Because very many molecules hit the piston at one time, the knocking is felt more as a steady push. The force of this push is determined by the speed, weight, and number of the molecules.

If the forces on both sides of the piston are identical the piston does not move, but if the force inside the cylinder is greater the piston will be moved, in turn moving the car or whatever the pistons are connected to.

The force increases with the increasing speed of the molecules, caused either by heating the gas externally or burning a fuel directly inside the cylinder (as described above).

Part II

Regional Foci and Pressure Points

From the mid to late 1990s the hunt intensified for "new" or more secure supply sources of oil and natural gas outside the Middle East, together with a frenzied search for pipeline routes or transport infrastructures to bring these hoped-for supplies to the big consumer markets of North America, Asia and Europe. For oil, however, the situation is particularly stark, simply because we are now so close to Peak Oil, as shown by the dramatic figures and charts in the chapters by Campbell and Korpela in the previous part. Because of this, the industry of denial has beat the drum about vast new oil reserves just waiting to be found in both likely and unlikely corners of the world. Some are in Iraq, now under US control and therefore no longer "unstable." Almost immediately after Baghdad fell the world's press carried reports of its western desert hiding vast oil reserves, and some experts even suggested there could be large amounts under the palaces of ex-president Saddam Hussein. Mostly these exaggerated and distorted reports of "huge new reserves" concern areas far from the Middle East, such as West Africa, the southern Atlantic and Gulf of Mexico. Usually, and not long after, these reports are denied and the propaganda is revealed for what it is by simple facts. Perhaps the simplest is that the official energy watchdogs of the big consumer nations, the OECD's IEA and the US EIA, together with the major oil companies, all periodically and regularly revise oil and gas demand growth figures for 2002–03 upwards. To obtain balanced supply and demand, these agencies and companies cannot call on fantasy oil from under the Atlantic Ocean, but instead increase periodic estimates of "call on OPEC." This phrase carries a lot of symbolic meaning. The most basic is that OPEC producers are "the suppliers of last resort," always taken as being able to supply oil at a price the market might pay. Another key myth is that OPEC producers are not affected by depletion, rising production costs or domestic consumption needs outstripping production. The other category of exporters, "non-OPEC suppliers," essentially means Russia, Norway and Mexico, together with a gaggle of small producers mostly in Africa but also including Brunei, Oman, Colombia and a few others, who currently produce more than their domestic markets can consume. For gas alone, the focus is essentially on four areas – the two biggest regional consumer areas of North America and Europe, and the hoped-for massive new producer, Central Asia, together with the world's biggest single gas producer and exporter, Russia.

This book focuses on the near-term certainty of Peak Oil. As already discussed, a host of telltale facts and indicators complete the arguments set out by Korpela, Campbell and myself. More support for the "depletionist" argument is found in a swath of news and information sites on the Internet (see Notes). Peak Oil and Peak Gas are linked and interdependent. This is proved by the arguments set out by the OECD IEA in its "World Energy Outlook" series. In recent years this series has set out policy and arguments as to why gas must take the strains that will be caused by the *possible* slowing of world oil production growth. In other words, as we run out of oil we can ramp up gas production and everything will be fine. Through 2002 and 2003 in the US, however, the impending natural gas "cliff" in domestic supplies percolated up from the production areas and analysts' studies. The starkness of the "cliff" even led Federal Reserve chairman Greenspan, in June 2003, to warn the House Energy and Commerce Committee that short supplies of natural gas could contribute to what he called "erosion of the economy." Greenspan talked of US gas prices perhaps rising to US$7 per million BTU (equivalent to oil at US$41 per barrel) and staying there through winter, and called for recognition of "the potentially important role that liquefied natural gas [LNG] could play in American energy imports." Here we return to the world context of Peak Gas coming at least ten years after Peak Oil, with world gas supplies being increased rapidly. Both pipeline gas and LNG will have to "take the strain" as oil begins to fade. As for the US, its peak of domestic oil production was in 1970, as predicted by M. King Hubbert, and its peak of domestic gas production is occurring now (or perhaps was in 2001). US domestic gas production will continue to fall, and LNG will certainly be needed to bridge the widening gap. This is a long and complex subject and the meat of endless conferences and meetings, reports and policy papers – but essentially LNG is much more costly than pipeline gas, currently providing below 2 per cent of US gas supplies, and the actual rate of US domestic gas depletion will be critical in deciding when – rather than if – gas prices will increase faster than oil prices in the US, with an immediate effect on world prices. At certain times in 2003, daily traded gas futures peaked at well above the key or "panic level" of US$7 per million BTU, and some US analysts believe "spikes" of US$10 per million BTU may occur. Europe also faces the certainty of indigenous gas production (mainly from the North Sea and Holland) falling, but at a much faster rate than US domestic production. The planned response is a vast growth in pipeline gas imports from Algeria and Libya to the south, and Russia to the east, with or without extension of the Russian gas grid to accommodate Central Asian gas.

In the case of the US, the cover for the coming gap caused by domestic gas supply depletion and increasing demand relies heavily on LNG. Europe relies even more heavily on OPEC pipeline gas suppliers in North Africa, and pipeline gas from Russia. In both cases dramatic growth is projected, with huge investment in both LNG and pipeline infrastructures. The major problem is that little or nothing has begun to move – the huge industry of studies, reports, conferences and meetings has no counterpart in action on the ground, ensuring future price spikes and supply panics. The missing element is prices – gas, like oil, is a traded commodity. Any oversupply and the projected or actual price spikes turn into price crashes. Investments running at tens or hundreds of billions of dollars or euros simply cannot be financed in the face of such uncertainty. So players "wait and see." Those most interested in seeing development move – energy companies, pipeline companies, big gas consumers (like fertilizer and pharmaceuticals producers) and national governments able to consult dozens of "confidential" and other reports on the certain energy supply crunch – have tried to accelerate action. One clear and even laughable case is Caspian region oil and gas.

Rightly titling this region's role in the supply of oil "The Caspian Chimera," Campbell explains how the US government, and other players interested in seeing action in this region, thought to contain huge amounts of oil, simply played up the reserve numbers on a "Who-can-exaggerate-most?" basis of excited and enthusiastic communiqués and presentations. At one stage in the 2000–01 period (and not long before the US invasion and occupation of Afghanistan yielded a body count on US estimates of about 4,500 dead (about 90 per cent civilian), followed by quick participation in "peacekeeping" by other big gas and oil importer nations like Germany and several East European countries) Caspian oil "experts" reached their crescendo. Their estimates peaked in that period with the idea that surely some 200Gb (about three quarters of Saudi Arabia's official proven reserves estimate) would be found in the Caspian area. Campbell considers that the real figure may be around 30Gb, or about 1.5 times what remains in the North Sea after about 20 years' production. In addition, much of the oil that has been found is contaminated with sulfur and heavy metals, is difficult to produce, and will need very costly infrastructure development in the form of refineries and/or pipelines to export no more than perhaps 2.5 million barrels/day on a sustained basis. This is less than two years' *growth* of world demand. The real Caspian, or in fact Central Asian prize is gas only, with possible undiscovered "stranded" gas deposits being world class; but – as with oil – the production and transport infrastructure needs are daunting. From almost the day the Bush administration "discovered" the link between the Taliban and Osama bin Laden, and wags gave the name "Pipelinistan" to the now militarily occupied Afghanistan, the International Community decided to set up shop and install troops. All pipeline projects in Afghanistan are on hold and will likely remain that way.

Another war-wracked and economically ruined area of the globe – West Africa – has taken the limelight as a possible savior for the oil-greedy and oil-wasteful rich world. My chapter "Dark Continent" returns to the same basic argument found throughout this book: too little, too late. Real solutions to the Final Energy Crisis are not on the supply side, or at least not on the *fossil energy* supply side, and not only because of accelerating depletion and the climate change caused by even the present level of carbon-containing fuel burn, but because the throwaway economy and society of the Western world is a doomed experiment in defiance of almost every resource, energy, environment and planetary limit that can be described, analyzed and studied. Our present crop of politicos naturally do not see things that way, and the "Joe and Jill Six Pack" voters, when or if able to think about the subject, would loudly proclaim that some crop of Einsteins will instantly cobble together a host of quick fixes in their garages or laboratories, just for Joe and Jill! However, at least their oil greed has resulted in some steps being taken to calm down and resolve the pan-African war that has claimed the lives of tens of millions since the 1980s. The latest recruit in the search for offshore oil to send to the US, Europe and Asia is Liberia, where the oil prospecting effort will at least require a semblance of civil calm and a pause in the constant civil war.

This section also contains a chapter by the late Mark Jones, who shortly before his death in 2003 was an adviser to LUKoil (Russia) in drawing up oil and gas contracts. Mark's chapter is in some ways the most urgent because of its warning of what the countdown to Peak Oil will cause in the arena of Great Power rivalry. China and the US are sure and certain rivals. The Final Energy Crisis will unquestionably reveal this fundamental rivalry, the need for conflict between these Titans dancing on the greasy slopes of the oil and gas cliff. A few figures explain all. China presently consumes about 1.6 barrels/person/year, *increasing at 6 per cent or more each year*, while the US, champion of oil resource destruction and profligate fossil energy consumption, has attained the probable absolute peak of about 25.6 barrels/person/year, with demand growth being episodic, as in 1999–2003, and at low annual average growth rates. China's oil imports are growing at constant and staggering rates, in the region of 30–50 per cent per year, as its domestic production first flattens (in the 1990–98 period) then starts to significantly decline (from 1999). China is only following the path trodden by the US since 1970–71, when the US attained its Hubbert Peak of domestic oil production. Remaining world reserves – especially oil but also gas – are concentrated in the Middle East and Central Asia, together containing at least 55 per cent of remaining oil and around 45 per cent of remaining natural gas. Since 2001 the US has installed military camps, resources and even the militarized colony of Iraq in the Middle East and Caspian region. Conflict between China and the US, sparked directly by oil and gas or triggered by economic crisis due to runaway oil and gas price rises, is far from unlikely and could require a build-up of only a few months, depending on the scenario.

The chapter by Marc Saint Aroman and André Crouzet returns us to the fickle hope of massive nuclear power plant construction as a solution enabling a few more years – perhaps a decade or two – of high-energy living and throwaway consumption for the world's rich democracies and those countries in Asia that have so successfully imported this model. First, we can note that since the 1990s nuclear power has virtually stopped growing. In the US and Europe the number of plants built since 1985 can be counted on the fingers of one hand, while those being programmed for "decommissioning" need the fingers of several hands. As this important chapter notes, nuclear power is part of the "talkfest" that hides hard-to-reconcile world views. The well funded and official pro-nuclear lobby always forgets to admit and accept that nuclear power – in the old world democracies – is a simple adjunct to and outgrowth from nuclear weapons development, and that so-called "civil" nuclear power is impossible without hidden subsidies, which are naturally kept from public awareness. For France (with about 85 per cent of its electricity of nuclear origin) this situation is very clear. Elsewhere, it is cloudy and full of the most extreme risks for life on this planet. India, Pakistan and North Korea have all clearly proved that civil and military nuclear power are one and the same thing *when the political decision is made.* They have also proved that civil power-generating reactors can come before the military phase, rather than after it as in France, the US, China, Russia, the UK and Israel. Many other nations possess nuclear power plants – any one of them can produce nuclear weapons in a few months, if its political masters so choose. Thankfully, nuclear power is so costly, so dependent on cheap energy subsidies from oil and gas (for fuel production and reprocessing), and so dangerous, that construction is almost nil. Any decision to reactivate nuclear power plant construction on a massive scale is almost certainly doomed because of long lead-times and enormous costs, in the context of the very rapid wipeout of cheap oil resources that will so compress electric power demand that any need for the nuclear "solution" will disappear. This we can hope.

5

The Caspian Chimera

Colin J. Campbell

The Caspian is one of the most ancient oil provinces of the world. The Zoroastrians of antiquity worshipped the eternal flames of Baku, which were smoldering hydrocarbon source rocks and gas seepages. F.N. Semyenov drilled a well there in 1840, operating under a concession granted by the Tsar of Russia, eleven years before the self-styled Colonel Drake drilled his well at Titusville, Pennsylvania, which is commonly taken to mark the start of the modern oil industry.

Geographically, it is a salt-water inland sea or lake covering about 375,000 square kilometers, bordered by the Elburz Mountains of Iran to the south and the Caucasus to the northwest. The Volga River flows into it from the north, forming a large delta near Astrakhan, but evaporation is sufficient to counter the influx, leaving it some 30 meters below world sea level. It is flanked to the north by Russia itself, followed clockwise by Kazakhstan, Turkmenistan, Iran, and Azerbaijan. The three "-stans" gained independence following the fall of the Soviets in 1991. Dagestan and Chechnya, which are still Moslem provinces of Russia on the shores of the Caspian, are still seeking their independence, in a vicious campaign attended by many acts of terror. Under international law, ownership of the offshore mineral rights depends on whether it is deemed a lake or a sea. In the case of the lake, they belong jointly to the contiguous countries, whereas in the case of a sea they are divided up by median lines. The matter, which is no small issue, has yet to be fully resolved, but it seems in practice to be moving in the direction of the latter formula. It is worth noting here that Tehran, the capital of Iran, lies only 100km from the Caspian shore, so its role in the future of the region cannot be ignored.

In geological terms, it is made up of several diverse provinces. To the south there lies a deep Tertiary basin in the fore-deep of the Elburz Mountains. It is followed to the north by the proto-delta of the Volga that runs across the Caspian as a fairly narrow belt from

Azerbaijan to Turkmenistan. That gives way to a Mesozoic basin, running out of Kazakhstan, which in turn adjoins the southern part of a large Paleozoic basin, known as the pre-Caspian basin, whose axis lies to the north of the Caspian.

Early oil activities were concentrated on the Aspheron Peninsula of Azerbaijan, around the town of Baku on the proto-delta of the Volga. Oil and gas, generated in lower Tertiary deltaic sediments, has migrated upwards, mainly along fault-planes, to accumulate in a thick sequence of Miocene and Pliocene sandstone reservoirs at fairly shallow depths. A peculiar feature is the so-called mud-volcano, in which gas seepages carry mud to the surface giving volcano-like features, several hundred meters high, which occasionally ignite and explode. Extensions of this same geological province extend northwards into Chechnya, where many people still make a living refining oil from shallow wells and seepages in primitive, dangerous and very polluting home-made stills.

Baku was one of the great world oil centers during the late nineteenth century. The Nobel brothers of Sweden held a dominant stake, later joined by the Shell Oil and Rothschild interests that financed a pipeline to the Black Sea. No less a figure than Joseph Stalin had his early experiences in Baku as a workers' leader facing the appalling operating conditions of the early oilfields, a ready breeding ground for revolutionary ideas.

The Soviets were very efficient explorers, as they were able to approach their task in a scientific manner, being able to drill holes to gather critical information, whereas their Western counterparts had to pretend that every borehole had a good chance of finding oil. In the years following World War II, they brought in the major producing provinces of the Union, finding most of the giant fields within them. Baku was by now a mature province of secondary importance, although work continued to develop secondary prospects and begin to chase extensions offshore from platforms. The Soviet Union had ample onshore supplies, which meant that it had no particular incentive to invest in offshore drilling equipment. The Caspian itself was therefore largely left fallow, although the borderlands were thoroughly investigated. Of particular importance was the discovery of the Tengiz Field in 1979 in the prolific pre-Caspian basin of Kazakhstan, only some 70km from the shore. Silurian source-rocks had charged a carboniferous reef reservoir at a depth of about 4,500 meters beneath an effective seal of Permian salt. Initial estimates suggested a potential of about 6Gb, but the problem was that the oil has a sulfur content of as much as 16 per cent, calling for high-quality steel pipe and equipment, not then available to the Soviets. Development was accordingly postponed.

The fall of the Soviet regime in 1991 opened the region to Western investment. The oil industry in particular was enthusiastic that here might be a "new frontier" to offer them another lease on life, having effectively lost the Middle East through expropriation, and having thoroughly explored the rest of the accessible world. A glance at the map of the unexplored Caspian Sea surrounded by oilfields was enough to capture the attention of Western strategists, especially in Washington, who began to hope that in the Caspian they could find an escape from the stranglehold of the Middle East in their desperate quest for access to a foreign oil supply. These notions and ideas soon gained a momentum of their own, far removed from any thorough scientific analysis. There were many motives to exaggerate the prize, as strategists sought to shift foreign policy and mobilize military capability. Before long the Caspian had won the image of being a second Saudi Arabia, floating on oil.

A second look at the map reveals that it is not easy to get the oil out of this landlocked area, remote from Western markets. But this was manna from heaven for various geopolitical "experts," who could now dedicate their think-tank efforts to designing devious strategies for controlling the transit countries and building pipelines, of course taking the claimed geological potential for granted. As many as eleven schemes were considered, each with different obstacles. The obvious route was through Iran, but this would have given Tehran a critical control of future US supply, which was not thought desirable. Another was to the Black Sea, for shipment through the Bosporus in tankers, but the Turks objected that this would be an environmental catastrophe waiting to happen. Existing pipelines through Russia could be used and expanded, but that gave the Russians critical control. The Chinese too, who recognize their desperate dependence on growing imports, entered the scene with a proposal for a pipeline in their direction. Then there was Afghanistan, and a proposal by the American company, Unocal, for a gas pipeline from Turkmenistan to the Indian subcontinent. Another heroic idea was to pipe it to the Black Sea for trans-shipment to Bulgaria, for whom environmental issues are not a particular priority, and then into another pipeline constructed through the Balkans to the Adriatic coast of Albania, passing through Kosovo, again a route not without its hazards. The preferred route at the time of writing appears to be overland through Azerbaijan, Georgia and Turkey, which would be no mean undertaking, having to cross high mountain ranges occupied by disaffected Kurds and others, who might find it a ready target.

In any event, the US began to establish and expand its military presence throughout the Caspian region with various bases and military "aid" projects in the nearby countries, and later toppled the Afghan government after a short campaign of bombing supported by ground forces of the Northern Alliance. The new president, Hamid Karzai, was himself associated with the Unocal pipeline project, directly reporting to Dick Cheney, and this project has now been resurrected. Meanwhile, enterprising Irish entrepreneurs import Caspian oil to Iran, re-exporting Iranian oil in exchange. All of this can be seen as a kind of replay of the so-called "Great Game" where, in the nineteenth century, various Western powers and Russia vied with each other for influence in Central Asia. However exciting this may be, it now begins to looks as if the Caspian may not live up to expectation, as ten years of exploration and development by Western companies reveal its real and modest potentials.

BP took a pioneering role with Statoil, its Norwegian partner, when the Caspian opened. Interest was at first aimed at the offshore extensions of the Baku trend, where a number of prospects, already identified by the Soviets, were successfully tested, leading to the development of the Azeri, Chirag and Guneshli fields. Some 17 "wildcats," as exploration boreholes are colorfully termed, have been drilled since 1992, finding some 3Gb of oil, which while useful, is not enough to have any particular world significance. BP also investigated the Tertiary deep to the south, finding the Shah Deniz, a gas-condensate field. Evidently high temperatures on deep burial have broken down the oil into a gas, containing a high dissolved liquid content, as might be expected. This area verges on waters claimed by Iran, and seismic surveys have been halted by Iranian gunboats. It is significant that Russia's LUKoil, which was a partner in the Azeri-Chirag-Guneshli fields, has decided to sell out to a Japanese company desperate for access to oil; meanwhile Exxon-Mobil has withdrawn from Azerbaijan altogether. Evidence to date suggests that Azerbaijan has reserves of about 12Gb, and since the larger fields are almost always found first it is unlikely that new exploration will bring the total to more than about 15Gb, if that.

Kazakhstan also soon attracted serious interest. The American company Chevron (now Chevron-Texaco), together with Exxon-Mobil, agreed to develop the Tengiz Field. It faced many operating and technical challenges, but has managed to build production to about 250,000 barrels/day, which is exported partly by rail and partly through the Russian pipeline system. One of the problems has been the disposal of the massive amounts of sulfur that have to be removed from the oil by processing. Plans to increase production by 480,000 barrels/day by 2005 have now been shelved, adding another nail to the "Caspian bonanza" coffin.

The greatest interest of all, however, attached to a giant prospect, termed Kashagan, which was identified in the shallow waters of the northern Caspian off Kazakhstan. Like Tengiz, it relied on a high-sulfur Silurian source, deep carboniferous carbonate reservoirs and Permian salt seal. It had a huge upside thanks to its sheer size, offering a certain potential to become perhaps the world's largest oilfield. Jack Grynberg, the well-known New York promoter, managed to strike a deal with the Kazakh president, leading to the entry of a largely European consortium, comprising BP-Statoil, the Italian company Agip, British Gas, the French company Total (now Total-Fina-Elf), and minor American interests. Grynberg retained for himself a so-called "overriding" royalty. But the initial enthusiasm waned when the companies began to get into the details. In geological terms, there were uncertainties whether the reservoir would be one large platform, or would turn out to be made up of individual reefs separated one from another by rocks lacking porosity and permeability, as experience from Tengiz would suggest. Seismic surveys showed that the integrity of the salt seal was weak in parts of the structure. The companies also found that they faced monumental operating challenges: the waters were shallow, making it difficult to bring in and position equipment, while also posing environmental threats to the breeding grounds for sturgeon shoals supporting the Russian caviar fisheries. If that was not enough, a gruesome, chilling wind blows in winter covering everything in ice. Nevertheless, the companies have succeeded, at astronomic cost, in drilling three wildcats, on what presumably are the most favorable parts of the prospect, announcing that they had found between nine and 13Gb. BP-Statoil decided to withdraw, exposing themselves to a lawsuit filed by Mr. Grynberg, who was not pleased to miss his overriding royalty. This is another big nail in the coffin, although the remaining companies, now led by Agip, soldier on.

In addition to these main projects, the Russians themselves have made a 2Gb discovery in the northwest part of the Caspian, and

Turkmenistan has announced an oil discovery of uncertain size off its mainly gas-prone territory.

In short, it has now become clear that the offshore Caspian has been a great disappointment. Exactly how much of a disappointment is hard to say, as the oil statistics are even more unreliable than usual. Total reserves for the offshore probably stand at about 25Gb, with new exploration offering potential for perhaps another five, a good deal less than the 44Gb Mean estimate proposed in a study by the USGS in 2000 (and vastly less than oft-publicized wild estimates, extending up to "200 billion barrels" in the US and European press, through 2000–01).

Offshore production today is mainly confined to Azerbaijan, where it probably stands at about 250,000 barrels/day. Given the withdrawal of the major companies, the monumental technical and operating challenges, the uncertain contractual regime and export pipeline obstacles, it is difficult to be sanguine about the future rise of production. Realistically, it seems doubtful if it will be possible to reach a maximum of more than, say, about 1.5 million barrels/day in ten to 15 years' time. If this plateau was achieved, and it would be effectively constrained by pipeline capacity, it might last another ten years before the onset of gradual decline at the then depletion rate.

The US currently imports some 11.6 million barrels/day, approximately 60 per cent of its consumption. If demand were held static by recession or government policy, imports would still have risen to about 17 million barrels/day by 2015, in the face of the continued decline of indigenous supply. About 10 per cent of its needs could come from the Caspian offshore, in the unlikely event that it was able to have exclusive call upon it. And even that would last only for a few years.

The foregoing discussion relates to the offshore Caspian, which seemed to be a particularly promising area, not having been explored by the Soviets. The surrounding onshore territories were thoroughly explored, so that most of the prospective basins and the larger fields within them have already been identified. There is naturally scope for more exploration and development, leading to production growth in the future, but that is another story. It is very evident that the Caspian has proved a chimera, dashing hopes that it would lessen US dependency on the Middle East. This realization perhaps explains in part why it now turns its guns on Iraq. There is at the same time a serious lesson to be learned: all that glitters is not gold. When the dust settles, Iraq may also be found to be able to offer less than was at first hoped, nature being immune to military intervention.

6

Dark Continent, Black Gold

Andrew McKillop

WASHINGTON, September 19, 2002 – Africa, the neglected stepchild of American diplomacy, is rising in strategic importance to Washington policy makers, and one word sums up the reason: oil. Within the next decade, recently discovered offshore reserves are expected to enable West Africa to outproduce the North Sea's oil rigs and capture as much as 25 percent of America's oil-import market.

New York Times, September 19, 2002: "In Quietly Courting Africa, US Likes the Dowry: Oil," by James Dao.

Through 2001 and 2002 a flurry of initiatives, state visits, studies, projects and actions underlined the growing importance of Africa for European, US, and Asian energy importers – if only to ensure some distribution of risk away from growing chances of battles, conflict and possible supply interruptions from the Middle East. Elsewhere in the oil pumping community, Venezuela has experienced brewing civil conflict, which one day could degenerate into civil war, prompting US invasion "to restore democracy." Nigeria also is menaced, in its role of cheap oil supplier to consumer civilization, and emerging Great Power rivalry for oil reach is a constant reminder of how coming Peak Oil presages future and *permanent* shortage.

The above extract from the *New York Times* is a typical, upbeat, distorted and exaggerated report boldly advancing the American expectation of West Africa rapidly becoming a major oil exporter region, producing more than the North Sea province (now falling, and likely to average about 5.9 million barrels/day in 2003 – see Part I). Typically, by forgetting to note that the *depletion rate* of North Sea production is at least as important as West African production capacity, such articles deflect attention from the crucial facts of West African oil and gas reserves. The actual reserve size is low. West Africa, both onshore and offshore, has so far proved reserves of mostly offshore, deepwater oil that amount to less than

2 per cent of confirmed world reserves. Recent West African oil and gas discoveries are in line with this difficult-access resource base; any additional production from the region in the 2003–10 period is unlikely to exceed about 2.5–3 million barrels/day – around 50 per cent of output from the North Sea. As the above article also forgot to mention, *total African oil production*, continent-wide and including established producers such as Algeria and Sudan, in North and East Africa, together with Angola, Chad, Equatorial Guinea, Nigeria and Gabon, is only about 7.7 million barrels/day.[1] This *total* production, continent-wide, was itself only equivalent to North Sea output in 2000 (see Part I). Because oil consumption by all nations of the African continent is so low, increased local consumption is the real short-term priority for African countries: any serious attempt at conventional economic development in Africa will probably lead to *reduced* oil export capacities.

Such facts do not worry sensation-seeking editors, and import-hungry OECD national leaders will keep their eyes riveted on the Dark Continent for one reason: because it is so dark. As the Light Pollution Institute and International Dark Sky Association websites[2] graphically show, so low are the levels of night-time, electricity-based artificial lighting in Africa relative to Europe, North America and Asia that the continent appears like an inky black hole. A few figures show why this is so: for countries in the Economic Community of West African States (ECOWAS), including Africa's second-biggest economy – oil-exporting Nigeria – average electricity consumption per capita is around 50kWh to 75kWh per year, compared to annual average European, Japanese and US consumption rates of around 3,500 to 5,000kWh per capita. The entire oil consumption of the continent's 50-plus countries and estimated 900 million inhabitants was around 2.6 million barrels/day in 1999 – less than that imported by Germany's 83 million inhabitants, and well below *one-third* what the US or the EU-15 countries import *each day*. With such magnificent economy of consumption – quite easy to achieve with typical GDP per capita figures around US$200/year – Africa can export a large proportion of its small production.

This situation may maintain itself, and may not. As many upbeat articles cheering on new production opportunities from non-OPEC countries add, much of any projected explosion in African production will be heavy offshore deepwater production, almost exclusively on the western side of the continent. While ignoring the towering costs of production in water depths exceeding 5,000 feet (see below), these upbeat articles will sometimes note that by being offshore, these installations can be sheltered from the many civil wars that have ravaged Africa since the 1980s. In the last ten years, especially in the former Zaire, and throughout West Africa, intensifying poverty – inflicted by the so-called International Community – together with population growth and ethnic tensions have combined to create a nearly constant Pan-African War. As a base for finding and producing cheap oil and gas, Africa this leaves much to be desired, but consumer civilization strategists and their military commanders must carry on.

THE CONTINENTAL ECONOMIC MELTDOWN

It would be no exaggeration to state that Africa has been more wracked, exploited, colonized and oppressed in the period from around 1985 to now than it ever was in the heroic times of white slavery and colonial war, either before or after the "Carve up of Africa" of the 1850–1900 period, or in any ensuing liberation war. This onslaught is by free-market forces, and with loans supplied under strict conditions by the World Bank and International Monetary Fund. During the hike in prices triggered by the first Oil Shock, and lasting through about 1975–82, international lenders fell over themselves to finance huge projects for minerals, metals and agrocommodity development throughout Africa. When the Thatcher–Reagan recession and slump "brought the world back to its senses," these loans became nearly impossible to repay, much like – for example – the US national debt which, in 30 minutes or less, will rise by several dozens of millions of dollars. The US trade deficit, running at about *US$50,000,000 per hour* in 2002, is in sharp contrast to structural adjustment conditions for the trade accounts that are imposed on African debtors. Squeaky-clean balanced budgets are policed to the last dollar, to the last kilo of food *not* imported, and not fed to tens of thousands of undernourished children. Black Africa's outstanding loans continue to this day to be topped up with interest rates on variable-rate loan schedules, set during the long years of extreme real interest rates through the 1980s. By 1985 Africa included a string of countries where more than 25 per cent of total export receipts for their agricultural and mineral or primary product exports, whose price levels had fallen far and fast from their 1975–82 highs, were needed simply to pay *interest due* on state-guaranteed loans.

Apart from some well-publicized debt forgiveness, sometimes operated on loan amounts that have snowballed through rising interest rates and accumulated arrears, the established procedure for dealing with Africa's unpayable debts is to consign these countries to the Club of Paris. This truly rich-man's club – except for the aberration of including downsized, impoverished Russia in its membership[3] – was set up for restructuring and consolidating loans after private banks threw in the towel on non-performing loans by African debtor countries. Essentially, the Club of Paris passes the ball to the World Bank and IMF, the lenders of last resort, who fix draconian conditions of further extended poverty as their price for bailing out poverty-wracked African borrowers, whose Sunset Commodity exports command such low prices they cannot repay their loans. The treatment applied – structural adjustment – is always accompanied by World Bank and IMF experts being jetted into the finance and planning ministries of the victim country; these well-paid and well-fed experts immediately ordering huge cuts in the number of public sector jobs to bring unemployment levels to at least 30 to 40 per cent. National assets, in the form of state companies, are immediately privatized. More often than not, these privatized companies are asset-stripped by US, European, Japanese, Chinese or Indian companies, and are then abandoned. Huge rises in the price of food, fuel, medicines and schooling always feature in structural adjustment. This further impoverishment of *already poor* countries is always nicely defended by various aseptic policy speeches and documents, but the applicability of or reason for *free-market pricing* in countries where often 50 to 75 per cent of the population lives *outside* the cash economy is hard to fathom.

Not surprisingly, this "shock treatment" has somber impacts on well-being. UNICEF gives estimates through the 1990s of *additional* infant mortality (five-to-nine-year age group) in Africa due to structural adjustment that run up to 1 million per year. This rightly named Belsen economics is now more widely known for what it is: the real "cutting edge" of neoliberalism, and its resource-supplying neocolonies. This is laughably described as making its victims "lean, mean and competitive," but it is hard to look at starving children and undernourished adults without public services alongside the neoliberal fantasy slogans which packaged them together so neatly.

Not surprisingly, the forced additional impoverishment of countries and communities already among the poorest in the world not only transformed black Africa's few oil and gas exporting countries into the most servile of price takers, but also prepared conditions for large-scale rebellions, massacres and wars. These conflicts in oil exporting countries such as Angola and Sudan were most "delicately" handled – arms were made readily accessible to those supporting cheap oil supplies for the West, while untold barbarity could proliferate from within the free-fire zones of these long-running mass killing sprees. Maintaining poverty in Africa was always a handy way to squeeze more "blood diamonds" to join the "blood oil and gas" that these warring price takers could be made to yield.

THE CHANGING CONTEXT

LAGOS, November 23 – Oil Production: West Africa May Overtake Saudi Arabia says Expert

The West African sub-region will in the near future, produce more barrels of oil per day than the current largest oil producer, Saudi Arabia, Mr. Van Dyke, President of Vanco Energy Corporation has predicted. Dyke said "Saudi Arabia currently produces eight million barrels per day while West Africa produces about 3.7 million barrels a day, primarily from Nigeria. West African production is expected to increase to 10 million barrels per day within five years, thus could soon be eclipsing Saudi Arabia's production."

Vanguard newspaper(Lagos) November 23, 2002

In the 2000–02 period, policy and attitudes of the so-called International Community towards black Africa changed rapidly because of oil and gas. As shown by the above, typically exaggerated report, black gold from the dark continent is a hope, or chimera reinforced by regional war in the Middle East no longer being a far-out, worst-case scenario. Exactly as with the Caspian, it is necessary wantonly to exaggerate reserve and production potential, but this mix of oil-greed and fear of losing supply sources has, almost directly, and very rapidly slowed, calmed and in some cases stopped several of the brush fire wars that were Africa's sole contribution to lightening its very dark night-time skies. In 2001, as Angola's off-shore oil prospects and development programs rapidly expanded, that "freedom fighter" so admired by President Ronald Reagan, and an honored visitor to the White House for fireside chats, the bloody Jonas Savimbi, was quietly shot one day, signaling the effective end of a 26-year civil war. In 2000–01 the long-running civil war in Chad – a three-horse race among Libyan, French and US-financed "players" – was brought to a rapid halt, at the very moment when a US$3.5 billion pipeline project received finance for the export of

Chad's recently discovered and developed oil reserves through Cameroon, for export to the US and Europe. And in Sudan, the long-running civil war between the Moslem north and Christian south has gradually had its venom removed, the free-fire zones being progressively pushed back from current oil production areas, and from those prospective areas for possible expansion of this country's tiny production. Elsewhere, the Algerian junta – controlling natural gas resources which supply about 35 per cent of EU imports – quietly continues its war with Islamic fundamentalists – a war that is "benignly" neglected by media and politicians alike in the "great democracies." For the time in which Algeria's gas reserves hold out, Algerian generals, who annulled the results of a 1992 election because their party was voted out, will very probably be able to count on state-of-the-art weaponry, training and diplomatic support from their backers – notably France and the US – normally so quick to denounce military juntas who "cancel" elections when they don't like the results.

Further, Africa has been condemned by the application of Belsen economics. Different powers apply their own and different methods. The US, for example, in 2002 announced that it would be stationing permanent troops in Equatorial Guinea, perhaps because this tiny country's tiny production of oil – little of which it consumes, due to abject poverty – is set to more than double to about 300,000 barrels/ day by 2004–05. In 2002, France established its SIGAfrique network with a 1.5 million euro grant to conduct national geological surveys of several West African countries, to store, provide security for and re-evaluate minerals and petroleum resource data, all to "safeguard the patrimony" of these countries. For many years the Swedish government has funded development efforts, now supported by the European Commission, to increase fuelwood burning in African countries that just might – dangerously – start to consume oil, and cause price increases for Europeans.

The rich world is wise – or blessed by geology – in focusing on Africa's offshore oil and gas, far from land and the danger of damage to installations. Africa's entire existence and survival is in fact threatened by the AIDS epidemic,[4] the certainty of increasing war and civil strife, and crushing poverty, itself increased by Belsen economics policies. Imagining that this "strategy" will enable the rich world to suck out cheap oil and gas and slow or stall the arrival of Peak Oil is dubious. It also starkly shows the level of immorality and inhumanity of the creators of this New World Order.

7

Battle of the Titans

Mark Jones

What is the nature of the current crisis? Where is it heading? What are the possible outcomes? The world geopolitical and economic system is holistic, and in many ways hyper-centralized and extremely fixed in how it behaves, but is composed of discrete elements with differing degrees of partial or relative autonomy. Different regions are subject to different dynamics. The rates of growth, or relative or absolute decline, differ between them. The system changes as a whole, but change also occurs in the existing equilibrium, and in balancing factors of regional economic and political power.

A.G. Frank, Immanuel Wallerstein and others have argued that the main trend in the world today is the decline of Anglo-Saxon hegemony and the re-ascent of Asia, and above all China. Legitimate questions arising include: Will this be a new American century, or is US hegemony as profoundly challenged as many now argue? Will China achieve regional hegemony, and is it capable of going on to true global hegemony? Or will China collapse when the sources of growth (easily identifiable and not the result of magic) fade, and underlying demographic, ecological, resource and inter-ethnic strains start to tell, possibly destroying the unitary Chinese socio-economic space, just as the USSR was destroyed by its failure in global competition?

A second group of questions: If there is a transition going on from the global hegemony of American to Chinese dominance, and a transition from the present Anglo-Saxon world system to a differently ordered world with Asia as its centre of gravity and propulsive dynamism, how will this transition occur and become effective? Is a world war thinkable, or can a peaceful transition take place from the Anglo-Saxon-centered world to a Sino-centric world? Can such a transition happen at all? Might the two systems abort in an endgame resource-war for declining oil reserves, through uncontrollable climate change, or other factors?

It should be borne in mind that the changeover from declining to ascending hegemony can happen – and has historically – not by means of war but with the consent and active participation of the declining power, which in some cases may even push forward the claims of the ascendant power, surrendering its own hegemonic status and freely giving up many strategic and economic possessions. There are examples of this from antiquity and in the Middle Ages, and the modern example is that of the surrender by Britain of its global-hegemonic status to an initially unwilling, isolationist US during the 1930s and 1940s. This process did to some degree contravene, or at any rate qualify, Lenin's thesis about inter-imperialist rivalry always leading to war, although arguably it was Lenin's own success in creating the USSR which caused this, by forcing the British to take a defeatist view of their own prospects. Is it thinkable that the US might surrender hegemony to China? Maybe it is more thinkable than we realize, once we canvass the alternatives, and once we look beyond the rabid posturings and imperialistic breast-beating of the US ruling class.

It is salutary to compare modern attitudes with those that were prevalent in Victorian Britain, during the period of unquestioned British global supremacy. Check out Rudyard Kipling or J.G. Farrell – the British were at least as sure of their manifest destiny, of their imperial, civilizing mission, and at least as arrogantly confident about an empire on which "the sun would never set," and about the racial superiority of their kind, as is the US today. Nevertheless, the time was not long before the British abandoned imperial pretensions, packed their colonial kit and left. Winston Churchill, in his desperate attempts to lever the US out of its isolationist neutrality in 1940, gave away many key strategic assets to Roosevelt, including not only the British and South African gold reserves, but the global network of island bases on which British imperial communications systems depended (this was the backbone of the information systems supporting the world markets of the time).

The middle decades of the last century might best be seen as an interregnum, during which time the declining and ascendant imperialist powers, Britain and the US, colluded and collaborated to fight rivals (Japan, Germany) and marginalize or contain rivals (Bolshevism). Once US hegemony was assured (by 1945) it was relatively straightforward to restructure the global system on a new, US-dominated basis, and to create the institutions and frameworks for global commerce and international law that secured unchallenged US hegemony. Except for two brief periods during the Korean and Vietnam wars, the USSR ceased to be a serious challenge after 1945, despite the much-vaunted menaces of the Cold War period.

The British surrendered their empire in pursuit of their own best interests, and indeed for national survival. But the psychological trauma and the bitter taint of defeat scarred a whole generation of people, not merely in the British ruling elites but among wider social classes with a sentimental interest in the empire. This included wide sections of the British working class. We can especially identify the so-called aristocracy of labor, which shared most of the racist assumptions behind the ideology of empire, and benefited materially from the so-called "social imperialism" of the military Keynesian/welfare-state reforms of the early postwar period, partly financed by, and riding on the back of, ebbing imperial wealth.

In its heyday the British Empire was more powerful and effective than the US empire has been or is today. The British not only moved populations around in huge numbers; they also moved plant and animal species. The British did more to shift and transplant alien flora and fauna from one continent to another, thus reconstructing whole ecosystems, than any other empire, although the Romans did a lot more in that sphere than most people realize. It takes a lot of arrogance and certainty to do the kinds of things the British ever-so-freely took it upon themselves to do. They reshaped whole continents, from Australasia through Africa to Latin America. Successive waves of emigration from the homeland created a whole English-speaking world, of which the US was at first only a subset and which it finally inherited, but had not created. The British ploughed their way through every precapitalist social formation they encountered, and either wiped it out or, through colonialism, totally reconstructed it. Yet despite (or because of) these grandiose achievements the British Empire, which seemed so enduring, was a short-lived thing. There is nothing to suggest that the seeming permanence of the US imperium should be any less fleeting. There is also no reason to suppose that sheer self-interest might not drive Americans themselves into a recognition that the price of domination, alone and unchallenged, is too high, and that an accommodation must therefore eventually be made with a rival who might one day become a successor. Since the 1939–45 war, the US has in fact made a practice of co-opting present and potential rivals into junior partnership. It has done this not only to Britain, but also to Germany (1960s), Japan (1970s and 1980s) and latterly even to Russia (from 1991).

The US is now functionally locked into Chinese industrial capitalism; the two states are in a tight embrace, performing a minuet which is part dance of death and part marriage of convenience.

Both states face many common problems, and each needs the other because of complementarities and useful asymmetries. However, the radical contradictions between them ensure that they are imperial rivals. And the balance of power between them appears to be changing. US economic and therefore military growth significantly declined from the later 1980s through to 2002. China has been growing faster, and has now entered a decisive phase of industrialization, where its industry is so diverse, deep, broadly based and synergistic that it appears to be crossing a threshold, and is emerging as the world's premier industrial power, eclipsing all others, with an R&D capability equivalent to that of the US or Japan. Many estimates indicate that Chinese industrial production will outstrip that of the US during the present decade. It seems clear that Asia as a whole is now poised on the cusp of precipitous changes, and that the US is now clearly the declining regional power, giving way to rising regional hegemony for China. It is surely arguable, or even likely, that China could become the dominant power in Asia without the need for war, and with the US being unable to prevent this. Once the economic facts are in place, can the geopolitical consequences be far behind? What can prevent the binding together and fusion of Chinese, Japanese and Taiwanese industry and capital, under Chinese hegemonic control?

Thus for the first time in its history, the US now faces the distinct possibility of the partial eclipse of its global power. I cannot predict what will happen, and I doubt whether anybody can know, but it is surely plausible that the US will be obliged to accept its strategic defeat with good grace, to accept a seismic change in its status and position, because the US is simply no longer powerful enough to resist. The rise of China to at least regional Asian ascendancy seems to be already in the script, an inevitable and unalterable outcome of present trends.

Lots of caveats and objections, surely, can be entered here. For one thing, it may really be true that US technological and military supremacy is now so great that it will be impossible for China ever to "effectuate" its latent regional supremacy, and take advantage of its potential power during some future world crisis. One cause, but also consequence of this would be the world economy entering a very protracted period of stagnation and decline. Also, China may be plagued by crises of national origin on one of a number of fronts: demography, environment, resource depletion, and so on.

A crucial indicator will be which state or region comes out of a depression first, and A.G. Frank rightly dwells on this. Under present circumstances, it seems unlikely that global bourses will recover very quickly, and the present bear market might be very prolonged. In fact, any protracted bourse crisis on Wall Street, with the index hitting even 4,000 points, will trigger panic equivalent to the 1929 crash, the post-1929 collapse of the fantasy paper economy initiating a six-year, worldwide depression in the real economy. Optimists argue that, even in the event of a paper-economy meltdown, this will not harm the real economy because mistakes made in the 1930s will not be repeated. That is: protectionism, futile and counterproductive attempts to balance budgets, monetarist rivalries, the deflationary gold standard, and so on. Wrong. They will be, and in any case even if enlightened Keynesian policies are in place, it may not help. Keynesian deficit spending to bolster the domestic economy has not helped Japan in the past decade of increasingly futile attempts to spend the country out of deflationary recession.

The same people who say that policy and regulatory improvements mean that another 1929-style crash cannot happen also often said there would never be another bear market, and that the New Economy was a "paradigm change." None other than Federal Reserve chairman Alan Greenspan himself became an eager convert to the neoliberal view that soaring stock-market index numbers were not, as he first thought, the result of "irrational exuberance" but represented a fundamental change in the economy and, in particular, a quantum shift in the rate of productivity growth. But we shall show that a slump can happen whatever kind of Keynesian demand management is attempted. Deficit spending does not overcome modern deflationary crises. So, as Wynne Godley argued in the London *Financial Times* in late 2002, a real and perhaps catastrophic slump, a real meltdown of the US economy in particular, is now a distinct possibility, even a probability. In that case, we might get a near-decade of mass unemployment in the capitalist heartlands, and a deepening pauperization of the peripheries. Yet this will take place on a geopolitical scene in which *the essence of the epoch is long-term competition for supremacy between China and the US.*

As Henry Liu argues, China abandoned its decades-long attempt at autarkic development during the Mao Zedong years. Instead, it has elected to join the world market and build a modern industrial system on the basis of export-led growth, rather than through self-sufficiency and import substitution. This is a highly dangerous strategy for both China and the world, although probably inevitable – China's leadership had no choice but to break out of the Maoist policy dead-end. It is dangerous for China because it entails an unsustainable commitment to growth through exports. Classical trade policy explains why overdependence on exports and capital inflows from abroad leads to further impoverishment of the masses, social tensions, and the inability to renew and develop essential social and economic infrastructures. At the same time, aggressive exporting helps destabilize the world economy. The counterpart to huge Chinese balance-of-payments surpluses are the increasingly unsustainable US trade and finance deficits. Additionally, aggressive exporting is inevitably deflationary. In effect, it pits the Chinese working class against the working classes of rival states. The question then becomes: Which of the rival, classic capitalist states is best able to raise its domestic exploitation rate? The winner will bankrupt its competitors. If successful, the strategy stands a good chance of destroying the bases of US world hegemony by destroying the US economy through endless deflationary down-spirals, starting with a savage cutback in overblown equity values. Chinese economic development policy is therefore – effectively, if not consciously – a policy of imperial rivalry and confrontation targeting the US.

The region which emerges first from a major and prolonged slump resulting from deflationary competition (and a slump is now surely on balance more likely than not) will be well placed to move out of mere regional dominance, and take its chances at becoming the unrivaled world superpower. If China survives the shocks and strains caused by a global slump in demand, then it is certainly well placed to emerge first from the subsequent trough. It is this line of thinking which leads me to argue that we probably are, as Frank says, in the throes of transition between hegemonies, and that indeed the present world economic crisis may itself be symptomatic of this crisis of transition, just as global collapse in the 1920s was symptomatic of the final decay of British power. China is more competitive, and this is the bottom line. Conventional or classic recovery for the capitalist world economy is likely to occur, at least through 2010–15, from any economic setback. The power with the underlying competitive edge will come out first. It will then be positioned to begin the process of institutional, legal and commercial restructuring to entrench its hegemony and ensure its dominance.

There is, however, one huge caveat to this whole line of argument. It does not take account of underlying, apolitical, extra-human and planetary limits, which will most certainly afflict the world system, and which mean we are entering a still more radical and decisive epoch of historical challenge, upheaval and transformation in everything from geopolitics to everyday life. As is indicated by the very title of this book, no geopolitical analysis can ignore the colossal threats posed by the storm of self-reinforcing crises – anthropogenic climate change, mass extinction of species, and destruction of the biosphere – all of which were enabled by fossil energy supplies, whose wipeout will be rapid. These factors form a vast backdrop to all world-system or economics-based considerations of inter-imperial rivalry à la Lenin. But before attempting to integrate this domain of issues into the discussion, let us consider again this central question: What if China emerges first from the probable imminent global economic slump?

This slump, if it occurs, will hit the US and Europe especially hard. The dollar will decline, industrial output will shrink, consumption levels and living standards will drop with incomes. The US economy will be first and hardest hit, notably because it is wildly unbalanced, extremely dependent on cheap energy, and equally dependent on the dollar's "reserve currency" hegemony, which enables its payments deficit to be ignored. Any US economic slump will inevitably bring with it a collapse in personal and public consumption levels. Obviously this collapse in US demand will hit major exporters, China above all. China will then have to find other outlets for its huge industrial output. This means exporting to other regions which may be doing little better than the US, such as Europe and Japan, India, Southeast Asia, Russia, and the Latin American countries – but perhaps firstly, countries which export oil, minerals, metals, and agrocommodities, which profit from higher prices for these items.

The major weakening of the dollar (if it happens) may take the US out of the current game for a very long time. We will then have a situation in which the US, too, must export its way out of trouble. Now we shall really see just how competitive the New Economy is, and how much US productivity really increased in the "dotcom" years. Under any hypothesis, however, no sane person would bet that the US can beat China at its own game. Walk around your house and mentally eliminate everything made in China, and see what's left. Now try finding anything with "Made in the USA" on it.

Once you strip out dollar hegemony and the advantages of being the global reserve currency (or "currency of last resort"), you are left with a very naked emperor. Except for weaponry, control of the skies and sea lanes, control and surveillance of world information networks, and a powerful propaganda machine, the US has few cards left. Take away dollar hegemony and you have just another regional economy with its fair share of internal problems (soil exhaustion, aquifer depletion, near-exhaustion of domestic oil and gas reserves, lack of alternative energy supplies, a polluted environment, poor infrastructures, badly designed and expensive-to-maintain urban environments). If the US has to compete on a level playing field with the rest of the world, then it may find that its urban infrastructure is just as uneconomic and unsustainable as was the Soviet Union's loss-making effort to base itself on the industrialization of the Urals and Siberia. The US currently uses twice as much energy and raw materials per capita as the EU-15 average, and more than ten times that of China. It is desperately uncompetitive. When the dollar has to be backed up by real values, US per capita GNP may fall by half in just a few years, as in the Great Depression. Under these conditions it is hard to see how the US can hope to maintain its global reach and present hegemonic position.

Since China faces similar global resource, energy and environmental challenges, and since the collapse of the world market must increase internal social instability, the Chinese regime will not wait around for the US to put the world to rights. It must produce – and export – or die, as outgoing President Zhou explained was the only choice for China during the 1997 Asian monetary crisis.

The two states who first came out of the Great Depression were Germany and the USSR – and each broke free through massive military spending. More recently, the Reagan "economic miracle" of the 1980s was largely driven by vast military spending, financed by government borrowing. Undoubtedly this is the first option large, militarized states consider when their leaders grope for reflation, lowered unemployment and re-emergence, with yet more power, on the world scene. A program of "military Keynesianism" is certainly an option for China, which is just beginning to expand and modernize its military massively. American defense spending is by comparison much less sustainable at present, let alone projected levels, because of skyrocketing trade and finance deficits, which will be intensified by a weakening dollar. Moreover, the extremely capital-intensive nature of US weapons programs means that defense spending does little to galvanize the wider economy. This is not yet true of China. Therefore, in a major world depression China could probably increase its national military spending dramatically, knowing that this will boost its economy while reinforcing its drive towards hegemony. The Chinese navy already carries out much gunboat diplomacy in the coastal states of Asia. The Chinese will surely seek to emerge from a depression through military and economic domination of the entire Asian region, by saturating markets and hegemonizing its skies, seas and data networks.

Sino-American joint or shared global hegemony may be the strategic compromise both states will entertain, for want of an alternative. Neither really wants war, but the resource and energy imperative may force the hand of either party – more likely that of the US. Under any hypothesis, however, the US will have declining hegemonic power, and China's will increase.

The last ten years have seen the greatest unforced capitulation in history – the uncontrolled implosion, or unconditional surrender, of the USSR, exploited by the biggest pyramid scheme in history (as Wynne Godley argues), all helping to create the biggest stock market bubble in history. The present bear market is not just a correction to that unprecedented human folly, unless you call Alaric the Hun's visit to ancient Rome a simple tourist's jaunt. This is surely the beginning of the end, not just of equity-culture, but of global Anglo-Saxon suzerainty. A.G. Frank was right: the pendulum is swinging back to Asia, but it is doing so under the final blow-out of the model of petro-capitalism.

If, on the other hand, we are set on a course of global war, which was the outcome for "classic" economic depressions before 1914, and again through 1929–36, then Americans have only a very small window of opportunity (like Hitler enjoyed in 1939) before their military advantage evaporates. This is perhaps the real cause of Bush's headlong rush to war. It is China they must pre-empt. The Islamic world, broken-backed as it was and remains, is not the problem. This will be a war for the survival not only of the American Century but of that cultural zone where people actually live – the "burbs" with an SUV in every drive – the pinnacle of ostentatious consumption, born and raised on cheap oil.

Even in his lifetime, Lenin recognized that his original assumptions about the necessity of inter-imperial warfare and the certainty of subsequent proletarian revolutions would have to be qualified in light of new realities. When the science-fiction writer H.G. Wells visited Lenin in his Kremlin office, Wells told Lenin that although he did not know what weapons the next war would be fought with, he was quite sure that the one after it would be fought with bows and arrows. Lenin did not disagree: it was already apparent, before the advent of nuclear weapons, that modern warfare imposed intolerable costs on civilization. It was this realization above all which prompted Lenin's notions of peaceful coexistence: What use would the proletarian revolution be, if it inherited a wasteland? The effects of civil war and war of intervention on the young Soviet Russia from 1919 to 1921 was almost as catastrophic as nuclear war. And the USSR never recovered from World War II, as Mark Harrison has shown in his admirable studies of Soviet economic development. The propensity of the bourgeois to overconsume and destroy the earth, rather than let the workers inherit it, presents a conundrum which neither Lenin nor any other revolutionary has successfully addressed. But in the twentieth century the bourgeoisie also learned a terrible lesson. While we should not assume that war between the US and China is inevitable, we can be sure that the dynamic of history shows that the decline of US hegemony *is* inevitable, and that China will be the beneficiary.

It is clear to both these powers, and any interested observer, that the keystone of US global power is Middle Eastern oil. This was true throughout the last century and is even truer today, as the US confronts a proven domestic oil reserve base of around 30Gb, and consumes 6.5Gb per year. The US energy crisis is both structural and accelerating. In the short-to-medium-term these energy supply difficulties might be met partly by conservation measures, because the phenomenal wastefulness of US society leaves much scope for saving. But this is not necessarily compatible with robust economic growth, despite the baying of Amory Lovins on the virtues of the unproven hydrogen economy. Above all, the US cannot afford to lose the economic race with China that at present it *is* losing. As I have already said, China's gross industrial output will probably exceed that of either the US, Japan, or Europe this decade. This will leave military control of Arabian oil as the remaining strategic asset – together with some military, intelligence and electronic technologies – to shore up the US global position. The US's pre-emptive move on Iraq will be largely designed to pre-empt China from asserting its power in the Middle East and becoming economically and strategically dominant there too. Can this US strategy succeed?

Looking ahead to the next 20 years, can Chinese hegemony consolidate itself? This would bring us back what may be termed, à la Chinoise, the Five Great Evils: anthropogenic climate change, mass extinction of species, destruction of the biosphere, resource depletion, and exhaustion of cheap energy supplies. But the Five Great Evils have no meaning by themselves. Rather, it is how people change in response to them that matters, and this is determined in the form and intensity of class struggle. At the moment conventional class struggle has almost no remaining political form, as mass societies around the world converge in a race to outwit, outpace or simply ignore, for a little longer, the gathering storm.

8

French Nuclear Power and the Global Market: An Economic Illusion

Marc Saint Aroman and André Crouzet

SO YOU THOUGHT THERE WOULD BE "OPEN DEBATE"?

Nicole Fontaine, France's Minister for Industry, signaled in late 2002 that France will lift all remaining restrictions on trading in French electricity and gas markets, under certain conditions, and before 2007. Ongoing negotiations to this end should be concluded by the end of November 2003.

Thus the "nuclear debate" announced by the Chirac-controlled and market-friendly government in January 2003 was preceded by an effective fixing of decisions, priorities and strategies for the French nuclear industry in 2002. From its earliest days, the French nuclear industry has been enabled and financed by the state, which continues to provide for this offspring. Yet the global market above all demands *economic performance* and the satisfaction of investors' desires, which in turn poses serious questions for the safety and security of nuclear installations, when the bottom line becomes the sole arbiter. Further, maintenance of the high-tech nuclear industry, and its fabulous resources of engineering capacity and manpower, can only divert resources and detract from the establishment of industries based on new, solar and renewable energy.

Open and full debate on energy strategies and policies in France is, as ever, delayed and diverted from the goal of obtaining clear responses to the question of creating new energy infrastructures, and to many legitimate questions concerning energy economics, geopolitics and the energy sector's environmental impacts, both in the short and long term, and resulting from our energy choices. Instead, the debate has focused on technical and industrial questions concerning how to save the French nuclear industry. In some ways this returns us to the very origins of France's long flirtation with the Friendly Atom, set in stone by governmental decree and imposed on the French in 1968, of course without the slightest democratic debate. At the time, the development of civil nuclear power was seen as a useful adjunct and support to ongoing military nuclear programs.

THE IMPOSSIBLE BOTTOM-LINE

It is in fact not possible to provide reliable and thorough financial, economic and trading accounts for the production of nuclear-origin electricity in France, for the simple reason that so much of the data are classified, secret, or deliberately modified.

Plutonium production for nuclear weapons was set as a national priority by General de Gaulle the moment World War II terminated, and the CEA (Atomic Energy Commission) was established to coordinate activity in this domain. By 1956 the site of Marcoule, in southern France, had three plutonium-producing reactors in operation, producing sufficient arms-grade plutonium for launching France's nuclear weapons program, whose first "test device" was exploded in Algeria in 1960. In a related development, EDF (Electricité de France – the French state electric power entity) undertook the construction of graphite-moderated reactors, all of the same design, with the declared intention of producing "atomic electricity." By 1965 six of these reactors were in service for EDF, bringing the total (military and civil) number of GCR-type, plutonium-yielding reactors to nine.[1]

Through the 1960s the Nuclear Origin Electricity Production Commission (PEON) gathered strength as a lobby within state planning and policy circles, and was rewarded in 1969, with the arrival in power of President Pompidou, by the decision to finance construction of PWRs (industry-standard reactors manufactured by GE and Westinghouse). Only from this period are figures of any kind available for the costs of nuclear installations. With the first Oil Shock of 1973–74, the French nuclear lobby was rewarded, perhaps beyond its fondest imaginings, by snowballing political feeling in favor of nuclear power for energy independence. In national prestige terms, also, France's decision to embark on a massive program of nuclear reactor construction was hailed as placing the country in the same technology playing field as the USSR and the US.

For the years 1945–68, as noted above, no figures of any kind can be given for expenditure on nuclear installations and facilities in France. For the period from 1968 to 2002, the cost of civil reactor construction – for a total of 58 separate reactor entities – is an estimated 153 billion euros. This ignores many aspects of the infrastructure and support for reactor construction and operation, and does not include "experimental" reactors, most notably the failed fast reactor "Superphenix," which for the years 1986–96 was estimated by the National Assembly spending review committee to have cost 7.7 billion euros, excluding decommissioning and other charges arising from its being taken out of service.

Thus the opacity of nuclear electricity accounting begins with the unknown, but evidently enormous costs of creating the initial infrastructures on which the industry depends. One thing is certain – EDF customers in France are subject to a levy of 15 per cent on their bills, this being termed a "provision towards decommissioning costs." The amounts generated by this levy are derisorily small when we consider the hundreds of installations that must be decommissioned, including uranium mines, nuclear fuel production sites, a total of 70 reactors (including "experimental" and military), "Superphenix" and its smaller cousin at Marcoule. In total there are *thousands* of tons of nuclear waste requiring storage and surveillance for *thousands* of years. The 15 per cent levy cannot even scratch the surface of the fantastic scale of the charges that are coming in the next few decades, but that will be borne far longer than the French nation has existed.

REPROCESSING: WHIPPED CREAM ON THE CHERRY PIE!

Under the pretext of being built to "recycle" nuclear materials (which are given the misleading name "fuel") the Cogema Cap de la Hague reprocessing installation entered service in 1967. As the then Minister of Industry Robert Galley later acknowledged, this plant was built with the objective of ensuring military plutonium supplies in the event of the Marcoule arms-grade reactor encountering operating problems, or suffering an accident, and cutting off military plutonium supplies.[2]

Reprocessing is fundamentally unprofitable, and has given rise to a litany of distorted and manipulated data. Some official and published reports provide insights into this chronic lack of nuclear profitability. A report in 1986 states: "Unfortunately recovered nuclear materials [at la Hague] do not have sufficient value to make reprocessing the profitable activity that we had initially hoped."[3] The July 2000 report by Charpin, Pellat and Dessus[4] shows that the costs to taxpayers of plutonium extracted from French civil nuclear wastes was between 170 and 290 million euros per ton. The 2002 report by Stoffaes[5] noted that: "It is necessary to clearly and completely break the linkage between nuclear origin electricity production and arms-grade military nuclear materials, by stopping the production of plutonium for bomb-making from civil nuclear wastes, and seeking long-term, acceptable solutions for the management of waste nuclear materials from civil reactors."[6] This being the case, how can continued reprocessing be rationalized or defended?

THE EVER-PRESENT THREAT OF GENERIC REACTOR FAULTS AND CATASTROPHIC ACCIDENTS

Official wisdom on nuclear industry problems and their sure and certain resolution by equally official solutions is a constant part of the scene. Despite this, while players in the nuclear industry try to cover themselves with respect to their peers and in relation to grave defects and weaknesses in current procedures and technology, their industry-only notes and reports, which the public never sees, paint a different picture. Much of this documentary material has the air of potential written evidence in case of litigation following major accidents. Continued aging of France's stock of nuclear installations throws up a continuous stream of newly discovered faults, and grave generic weaknesses. In relation to only civil engineering works, around 20 reactors currently in service face containment feature degradation and possible destruction through the location of their alternator sets. A clear majority of French reactors was designed and built with their operating level (height above sea level) set too low for present – much less increased – sea levels; over half of French nuclear reactors are now in flood-prone areas. In these cases there is only one real solution – the reactors must first be stopped, and probably decommissioned. The September 11, 2001, terrorist attacks revealed a potentially fatal weakness for France's 58 civil reactors – their protection features are exclusively designed to protect against core-related accidents and faults. No attention was given to the potential for external attack, leaving the country's reactors like tortoises without shells, because safety equipment and emergency cooling systems are located only outside, vulnerable to any terrorist!

Concerning reactor aging, the EDF regularly requests further time for replacing components and subassemblies that suffer accelerated aging and, for example, VPHC (vapor pressure head cracking or neutronic aging) syndrome. A wide range of components critical to safety are subject to aging, including pressure vessel covers, pressure heads, fuel rod sheathing, moderator rods and controls.

The Three Mile Island disaster of 1979 was not a human disaster, but was most certainly a financial catastrophe, provoking hasty modifications of most other reactors in the US, with a total cost estimated at close to that of all previous nuclear construction in the US, when downstream impacts (notably the freeze on all new reactor orders) are integrated. The Chernobyl catastrophe, after a long silence in the official media (including UN websites until January 2002) is now recognized as having caused economic losses of about 250 billion euros. To put this in perspective, this is at least twice the cost of all Soviet civil nuclear construction in the period 1954–90.[7] While French media, in the past few years, have admitted they were wrong to have announced in 1986 that fallout from Chernobyl stopped dead at the French frontier, they have yet to note that some 20 reactors were still under construction during Mitterand's regime, in 1986 when a certain Jacques Chirac was Prime Minister, and that the nuclear lobby pressed hard to commission these reactors into service.

MARKET LIBERALIZATION?

Liberalization of the electricity market has dealt a profound shock to the EDF. Price and cost analysis for electricity has begun to incorporate off-balance-sheet items that were formerly paid using state funds. In 2000 the management of the EDF called for a 30 per cent reduction in global costs, notably through flexible working hours, temporary employment contracts and other "liberal" employment practices such as massive out-sourcing and subcontracting. These changes have had major impacts on management–union relations; the primary workers' union in the French electricity industry, the CGT-Mines and Energy, in a notice to members calling for a demonstration on October 3, 2002, stated: "The breaking up of previous salaried staff structures, and constant pressure from management for reducing labor costs through massive outsourcing leads us to have serious doubts for the safety and security of installations."

To these contractual and labor relations issues the application of recent European Commission directives on workers' radiation exposure, setting thresholds 2.5 times lower than those previously used by the EDF, presents the EDF with major challenges for both costs and workers' safety. Time and cost constraints for replacing people with robots where radiation levels cannot be reduced throw up more constraints on the EDF's new ambition to enter the global market. The most recent internal audit by the EDF, presented to the National Assembly financial committee on September 18, 2002, announced that the EDF's accounts were edging into the red. Up to June 30, 2002, the EDF has lost to competitors one-third of the customers in France and Europe it had identified as prospects.

Further, the EDF's ambitious state-backed program of buyouts in Italy, Britain, Mexico, Brazil, Argentina and China has not produced the expected profits, and will certainly be unable to provide the level of financing needed for decommissioning national reactors, besides long-term management of waste. By 2006 the French government will hand down new and costly decisions on nuclear waste management.[8]

The EU Competition directorate, headed by Commissioner Mario Monti, has for many months targeted the EDF, and has moved to take legal action against it for unfair trading. One part of this legal action seeks to prevent the EDF from drawing down nuclear decommissioning funds to finance its buyout strategy. In its accounts, the EDF defines three types of provisions:

- For liabilities and charges, including nuclear waste reprocessing and plant decommissioning, currently 51.1 billion euros;
- For renewal of concessions on works, infrastructures and land which the company does not own (owned by local authorities), 20.7 billion euros;
- For retirement pension payments (counted together as an off-budget item), 42 billion euros.

The total for these provisions exceeds the EDF's net worth, with its own funds as of December 2001 standing at 13.7 billion euros, and debt at 22.2 billion euros.

EDF spokesman François Roussely, whose testimony to the National Assembly industrial production and trade committee in July 2002 announced that the company's accounts were edging towards meltdown, said in September 2002 that he was hoping for a cold winter to offer a single ray of hope for EDF sales and profits. Given the trend towards global warming, it appears that even the world climate is hostile towards the EDF's nuclear-based financial condition.

HOW DOES NUCLEAR POWER SHAPE UP TO FREE MARKET COMPETITION?

The TEPCO scandal in Japan (falsified documentation on nuclear materials for power reactors) only serves to demonstrate what had been long-term, standard practice in the Japanese nuclear industry, with its close links to Cogema and its Cap de la Hague reprocessing plant. The UK, whose Sellafield reprocessing plant has regularly and demonstrably falsified documentation, and which would have been bankrupted many years ago without government bail-outs, announced in December 2001 that reactor decommissioning costs of 56 billion euros would be borne by the state. One major objective of this financing program is yet again to save the BNFL Sellafield plant from bankruptcy. The reprocessing plant, because of document falsification, has lost several major Japanese contracts. Various financial "engineering" practices will be used by English administrators to transfer costs of the BNFL Sellafield plant to new shell companies and holdings,[9] with taxpayers' funds providing the financing. The privatized, nuclear-only UK electricity producer, British Energy, made the headlines in 2002 as it crawled from one restructuring plan, through bankruptcy, to another. From September 2002 to the first quarter of 2003 alone, government handouts to British Energy (supplying about 20 per cent of UK electricity) reached 650.8 million euros. One major cause of British Energy's staggering underperformance is what its directors call "overcheap" competition – electricity produced from coal, hydroelectricity, windmills, gas and even oil – in short, *any source except nuclear*.

WHAT ENERGY FUTURE FOR FRANCE?

The above "energy liberalization" policies and strategies, if applied by France, would lead to the same route of privatizing profits while taxpayers bear the costs of nuclear wastes and reactor decommissioning – which are essentially open ended. This doctrine better suits a kleptocracy than a democracy. The sirens of the global market, working on EDF's privatization from their tightly guarded offices of the Concorde Foundation in Paris, are calling for this "modernizing strategy" with mounting insistence.

All is grist to the mill for this "historic advance." As noted by O. Postel de Vinay, the editor of the monthly magazine *Research* in its December 2000 edition: "French nuclear advocates know that the greenhouse gas issue, to which they pay lip service, is a great marketing opportunity to seize on at this very moment." The real choice for France is either to hold the nuclear course, or, as many nations have already done, to leave it behind.

The French parliamentary report writers, Birraux and Le Déaut, recently noted of Germany's decision to abandon nuclear energy: "Given the strength of Germany's industrial capacity and the number of both domestic and international market opportunities for new and renewable sources of energy (NRSE), what might have initially appeared as economic suicide could in fact be the very reverse, the choice of a major strategic change at the right moment, and therefore the best possible decision."[10] The abrupt change has already led to the creation of around 35,000 jobs in the wind energy sector, and Germany's government has announced that up to 120,000 jobs will be created in this sector by 2010.

Conversely, if the French government continues to pour cash into the nuclear bucket as it has – for example through financing the Electric Power Reactor (EPR) (new generation) reactor construction program for which immediate requirements are for 3 billion euros – this will precipitate the continued waste of public finances. Continuing along this nuclear route will limit or exclude other energy choices, notably the NRSE programs and projects increasingly chosen in other European countries.

The current nuclear/non-nuclear energy mix offered by the French government is effectively a fake choice, partly because of the following:

1. France does not have the financial and technical resources to operate both nuclear and NRSE strategies at the same time. Costs and charges for maintenance and replacement of the country's current nuclear installations will swallow the bulk of resources, effectively eliminating NRSE choices. French policymakers are essentially nuclear die-hards, as shown by a typical reaction from political leader Mr. Bataille, who stated that the introduction of windmills for electricity production "would be a costly error." France has been lagging behind other countries in the development of NRSE and energy conservation, and is unlikely to catch up anytime soon.

2. Underlying France's nuclear strategy is the conception that electricity consumption must be increased to the highest possible levels.[11] Apart from thermodynamic efficiency and energy-economic considerations that go against continually raising

electricity's share of total energy consumption, the policy of "cheap" electricity is completely hostile to energy conservation and efficiency increases. Several other EU countries that have abandoned the nuclear route have done so on the basis that nuclear electricity is part of an overall policy that is anathema to energy conservation and efficiency.

CONCLUSION

As has been briefly noted in this review of French nuclear electricity, the most salient point is that it is *a financial aberration* from start to finish. The health and environmental effects of France's massive nuclear energy industry are increasingly grave and various. Enormous, in fact unquantified amounts of nuclear wastes, varying in toxicity and radiation levels, are simply dumped in rivers, ponds and waste tips, and certainly enter the food chain. No official recognition of this grave problem has yet been made. As one example of the "conditioned reflex" of French authorities in defense of nuclear energy, we can cite the work of Professor J.F. Viel. His findings of abnormally high leukemia rates in the Cap de la Hague region were published in the respected *British Medical Journal*. In response to Professor Viel's publication, the inquiry commission set up by French authorities officially found these high leukemia rates to be due to "random epidemiological factors."

On a worldwide basis, nuclear-origin electricity furnishes about 16 per cent of world usage and below 6 per cent of commercial energy. Nuclear-intensive programs like France's should be evaluated according to their costs and risks, relative to their small contribution.

A host of "externals" and "off-balance-sheet" costs and charges are screened away from the cost-per-kilowatt numbers that nuclear advocates like to brandish, when they gaily talk of world nuclear electricity production increasing 18-fold.[12] The range of these subsidies is enormous, from the iodide tablets given to anxious inhabitants around installations, emergency decontamination planning, nuclear crisis coordination exercises, new anti-terrorist provisions, and of course waste management and decommissioning. All these costs and charges are borne by tax-paying communities, as are insurance risks. Under current provisions, the EDF's insurable risk and maximum liability in the event of a "major" nuclear accident is set by agreement between the EDF, the French state and insurers at the derisory amount of 625 million euros. This can be compared to loss and damages arising from the TotalFinaElf nitrate factory explosion in Toulouse (September 2001), which are estimated at up to 1.7 billion euros. This major industrial accident, fortunately, did not render hundreds of square kilometers uninhabitable for hundreds of years, as the worst nuclear accidents do.

Democratic decision-makers and parliamentary representatives, if they have the power and courage, should take the only decision that is reasonable: they should set an exit date for all nuclear programs, in France and elsewhere. Nuclear accidents are unfortunately certain; the more aging reactors and installations, waste dumps and "civil" plutonium repositories that exist, the more catastrophic events of Chernobyl's magnitude and type will occur. Additionally, any nuclear facility of any kind is the largest, most lethal target for any determined terrorist.

Part III

False Solutions, Hopes and Fears

Preceding chapters have spelled out the case for Peak Oil, with Peak Gas not far behind. The initial changes that should be made have also been identified. In this section, C. Campbell provides the most concise summary of how his many decades of research and study as a petroleum geologist for several major oil corporations make him certain that Peak Oil is imminent. Of course, as we will soon be hearing in the media, "natural gas will save the day," and after that there will be the miracle of the Hydrogen Economy. But when the world turns the corner from the present path it will enter a long and accelerating downslide in cheap energy availability. Rather than waste time on so-called technological fixes, the inevitable future problem of food production should be addressed immediately. Transition to sustainable food production is at least as vital as the transition to a low-energy economy and society utilizing renewable energy and the long-term stocks of coal that exist. As Edward Goldsmith has explained, the transition to sustainable agriculture is totally feasible and practicable. However, no person could accuse Goldsmith of foolish confidence in the ability of our current mass consumers to move from burgers and French fries (now "Liberty" fries in the US) to grow-your-own organic health foods, even if those masses were willing. For the decision-making elites, the problem of will and intentions is yet more stark. To be a member of that elite you must have unbending confidence that you are right, and that business as usual is the only game plan for channeling the consumer masses to the fleeting nirvana of still more fossil-energy-based consumption. For many of those elites (excluding the Bush administration), a compromise on transition has been found in what might appear to be their surprising move to adopt the Kyoto Treaty.

My article on this arcane, confused, contradictory and surely ineffective set of good intentions — underpinned by genuine fears of real climate change — focuses on several of the Treaty's key targets: notably, reduction of carbon-containing emissions or "greenhouse gases," carbon sequestration (or the capture and removal of "excess carbon" from the atmosphere), and the CDM or "Clean Development Mechanism." This last needs to be highlighted, because it is nothing but a charter for lifestyle change, at first oriented to the developing countries, where establishing low and renewable energy usage, sustainable agriculture, and sustainable activities of all kinds, are the stated goals. In a real way, then, we can say that the architects of the Kyoto Treaty at least want to see the developing countries adopt a sustainable economy and lifestyle. This is an incredible mismatch, it must first be said very clearly. The energy-intensive economies and lifestyles of the rich, industrial urban societies, so heavily reliant on fossil fuels, are the *cause* of massive greenhouse gas emission. These economies and societies should first and foremost make the transition away from fossil fuels to a low-energy economy, habitat and lifestyles. Second, the real current aim of conventional economic development institutions and authorities is to accelerate the transition of developing countries *towards* the economy and lifestyles of the urban–industrial North. The Kyoto Treaty is therefore completely at odds with ongoing and existing development, which of course is explained away and fudged by this or that section, paragraph, rider or annex of the Treaty (the full document, with annexes, weighs about 22 kg). The CDM can be thought of as an agreeable experiment to try out in poor countries that may save a few grams of CO_2 per kWh or mega-joule in their national energy economies. In addition, there is no set, standard definition of what the CDM is, or in fact if there might be not one mechanism, but many. At present almost nothing is being said about the Kyoto Treaty, after its brief day in the sun around 2000–01, during which the Bush administration loudly denounced and rejected this very flexible set of nice intentions, good causes, and plainly ineffective measures.

Let us suppose renewable energy suddenly became the be-all and end-all of national energy policy and planning in the rich urban–industrial North. As the chapter by Ross McCluney (of the Florida Solar Energy Center) clearly explains, we are in fact confronted by limits of every kind. McCluney in no way ignores the fact that, in our advanced and sophisticated societies, we essentially *eat oil*, and starts with a brief examination of the intense fossil-energy dependence of industrial agriculture. As his chapter shows, we will very much have to cut our cloth as a function of what is available, accepting real limits on what renewable energy can or might do for us. The limits facing us could be divided into just two categories – those due to the generally low-intensity and variable nature of renewable energy flows, and those due to existing needs and desires for huge quantities of cheap energy at high power rates in the urban–industrial economy and lifestyle. Few motorists can appreciate, for example, that when they are filling their car in two or three minutes (or more for an oil-greedy four-wheel drive) the transfer rate is something like 10–20 megawatts, equivalent to *all* the energy that could be extracted or converted, with the highest-yielding equipment known, from the solar energy falling on several *square miles* in the middle of summer, in latitudes around 40° from the equator. However, they would immediately understand how privileged they are today if they possessed a hydrogen-fueled car, whose gas tank (costing tens of thousands of dollars or euros) would need perhaps 30 or 40 minutes to fill. This is due to energy transfer limits, and problems for the technologies developed and used for making energy easy to store, control and deliver. The fundamental thermodynamic reasons for this were discussed in Chapter 4 by Jacob Fisker, a particle physicist.

In no way would Ross McCluney conclude that solar and renewable energy is inapplicable, useless or a dead-end, but he does state very clearly that renewable energy is only a *part* of the solution to a set of interpenetrating and multiple challenges to the very existence of life on this planet, *caused* by the runaway population growth enabled by fossil fuels. We already do *eat* gas; we do *eat* oil – and are the descendants of those who *ate* coal. In the future, both oil and gas will be off the menu. In addition, that future is fast approaching for the key fossil fuel, oil, which is the prize of the Bush administration's foreign policy strategists and military planners, and those of China, as well as those of the EU leadership. When both oil and gas have effectively been relegated to minor fuels, we will have lost about 60–70 per cent of *all* the commercial energy we currently use in the advanced societies. This will happen in the next 30–40 years. Attempting to "make up the difference" with solar and renewable energy and some coal is a fool's quest. Decompressing the energy intensity of society, firstly in the high-energy, consumer societies of the North, is at least 50 per cent of the real solution, and will come either in a planned and organized way – requiring admission and acceptance of this need – or it will come through crisis, conflict and war.

Since the early 1980s there has been a dramatic transformation not of the economy – which now uses about 40–50 per cent more oil and gas than 20 years ago to crunch metals and extract commodities from the lithosphere and biosphere – but of society and culture in the older democracies of the urban, industrial North. Even the word "industrial" does not really apply, because of "outsourcing" and delocalization, as industry and manufacturing migrate to east, south and now western Asia, while the core nations or older democracies receive economic migrants and continue to consume industrial goods for a very large proportion of the planet. There is no single word for this transformation, but "apocalyptic" is one word that should be included in any set of terms, also including fear-laden, anomic, confused, stressed and aggressive. In addition to the "First Oil War" or Gulf War 1 in 1991, the founding event for this transformed society is the terrorist attack of September 11, 2001, an event with a rather clear Middle Eastern oil link, and the trigger for America's War on Terror, which directly led the US, with rather few allies, to invade and occupy Iraq. Already the oil prize from this invasion (about 10,000 dead, according to US and British official figures "to the nearest 500") is disappointing. Just like

Saddam's weapons of mass destruction, Iraq's oil reserves and capacities to produce were highly exaggerated, even imaginary – partly to attract loans to purchase conventional arms and finance the construction of palaces. Occupied Iraq may have extreme difficulties even re-attaining its pre-war oil export levels, partly because domestic consumption for a population now including about 175,000 foreign troops with a very high-energy "lifestyle" will most certainly increase rapidly. Before this war, uniquely due to oil greed, there were numerous large peace rallies in many old-world democracies, but since the event almost nothing stirs – no event can hold public attention more than a few weeks in a society where immediacy, distraction and futile pursuits are the very basis of an existence whose only contact with reality is to consume vast quantities of metals, plastic, glass, and chemicals – because we have no choice.

By 2035 we shall certainly have had many "choices" forced on us that could have been voluntarily and much more efficiently made right now. However, the consumer masses of the old-world North, glued to their TV screens and addicted to the latest gadget, gimmick or sensation, are even in late 2003 satiated with the War on Terror. Each atrocity must be bigger, better and more professional – probably with CIA-trained expert advisors – to merit the slick media campaigns built on communicating our complete dependence on those American-led military heroes who struggle against the hydra of Al Qaida, or whatever mysterious – of course Islamic – movement threatens our lifestyle today. The bin Laden industry is probably almost as well financed as the "limit denial industry," and very likely financed by the same organizations and powers. Maintaining the consumer masses in a childlike state of greed, terror and self-satisfaction is probably thought of, by our democratic leaders, as a great new tool for managing the future. The reason for this, I suggest in my chapter, is pure atavism. Our leaders of today, though they chirp Christianizing slogans, are much closer to the Roman Imperial war machine than to St. John and his Apocalypse. Imperial Rome was based on one thing only – continuous expansion. Our economic "empire" or "community" is based only on growth. Too many Western leaders today, especially in the US, are advised by their spin-doctors to show devotion to our Lord, make speeches in front of churches and have their children baptized. In the US the "cult thing" is the basis for hundreds of round-the-clock TV stations, and a vast political machine of influence-trading. Is it therefore any surprise that the apocalyptic fairy-tales of St. John are regular speech fodder for George W. Bush and his ilk, in what they call their war on fundamentalism, demagogy, intolerance and whatever – now lumped together as War on Terror?

It would therefore be ridiculous, even madness, to imagine that such leaderships can accept that limits apply to the growth economy and consumer society. Those bent on apocalyptic quests want fire, blood, destruction and conquest – and of course cheap oil, the lifeblood of what they call civilization. The consumer masses can of course feign complete ignorance and remain anxious that their children should go to university, study hard, and become *very* intelligent, or at least able to earn lots of money. However, this shield of numbness will do nothing to protect the consumer masses from change, and will make the coming transitions chaotic – unless and until there is massive change in public attitudes and political values. That is: culture change. This may come through economic crisis, if or when our leaderships admit and accept their policies and programs are *totally* unequipped to handle the real situation we face. This again, we can hope.

9

Oil and Troubled Waters

Colin J. Campbell

The millennium celebrations are fresh in our memories. Some gave thanks for the successful completion of the last millennium, while others welcomed in the New Year, the new century and the new millennium. Now, four years on, there is little cause for celebration, as the world finds itself at greater risk than at any time since the days preceding World War II. Indeed, a new war has already been declared, not against a particular country but against what is euphemistically called "terror." The attributes of the war are all too plain, from the dust of explosion to the screams of innocent victims, including those who suffer in acts of retribution, but the motives of the perpetrators are more obscure. The terrorist is not normally in it for money, but rather represents the dispossessed desperately seeking identity. The borderline between the terrorist and the freedom-fighter is a fine one. The terrorist too is often a proxy for distant sympathizers with a romantic notion of historical wrongs in the lands of their ancestors, whether we speak of Northern Ireland, Palestine or any one of the many divided communities around the world.

The Middle East is today the vortex of a storm that envelops us all, for the simple reason that it controls the world's supply of oil. This is not just another commodity obeying the normal laws of supply and demand as entrenched in economic theory and dogma, but is rather the lifeblood of the global economy, providing 40 per cent of all traded energy and 90 per cent of transport fuel. Furthermore, it is a finite resource for which no better substitute is in sight. Each tank filled with diesel fuel for a tractor is equivalent to a team of unpaid and unfed slaves, who would otherwise have had to hoe the field by hand. World population has risen six-fold during the first half of the age of oil, thanks in part to the use of fossil fuels.

The physical attributes of oil are now well understood from great advances in petroleum geology. It transpires that the bulk of the world's oil was formed in no more than two short epochs of extreme global warming, 90 and 145 million years ago, when prolific algal growths poisoned seas and lakes, producing the organic matter that became oil and gas through burial beneath younger sediments. The oil, once formed, was preserved only in certain tectonic settings, mainly in rifts that opened as the continents split apart. These geological facts underline how very finite it is.

A coal deposit may cover a large area, being mined only where the seams are thick or easy to access. If prices rise or costs fall, lower concentrations become viable under normal economic laws, but oil is different because it is a liquid not an ore, being concentrated by natural geological processes. It is either present in profitable abundance, or it is not there at all, as is confirmed by a glance at the oil map, showing how oilfields are concentrated in clusters separated one from another by vast, barren tracts that lack the necessary geological conditions. The world has now been so extensively explored that virtually all the sweet spots have been identified, save perhaps in certain deepwater and polar regions. About 40 per cent of the world's endowment of so-called conventional oil lies in just five countries surrounding the Persian Gulf.

We desperately need to know the size of the world's oil endowment in order to plan for the future. The information could be provided without particular technical challenge, but is denied to us through ambiguous definitions and lax reporting standards by an industry desperate to conceal the truth, and by governments who prefer not to know. We need a detective rather than a scientist to sift the reports for clues. Such a detective would likely present evidence to define *conventional* oil as the easy, cheap stuff, excluding oil from coal and shale, bitumen, extra-heavy oil, heavy oil, deepwater oil,

Table III.9.1 World Oil Endowment

Conventional Oil	World	Middle East
Past production	873	225
Future production from		
known fields	884	483
New fields	144	43
Total billion barrels	1900	751
Remaining billion barrels	1027	526

World consumption at 78 million barrels/day: 28.5 billion barrels per year.
Consumption at 82 million barrels/day (likely Peak Oil rate): 29.9 billion barrels per year.

polar oil and liquids derived from natural gas. It has supplied the most oil to-date, and will dominate all supply far into the future. He would go on to explain that *proved reserves*, as reported for financial purposes, means only *proved so far*, based on current wells, which may or may not say much about what the fields will eventually deliver over their full lives – a cause of great misunderstanding. He would stress the need to backdate the resultant revisions to obtain a valid discovery trend, by which to extrapolate future discovery. In short, he would provide the following best estimates in billions of barrels, recommending at the same time that they be generously rounded.

This means that the world has now consumed almost half of its endowment; and that half of what is left will have to come from Saudi Arabia, Iraq, Iran, Kuwait and Abu Dhabi. Two of these countries, Iran and Iraq, are declared enemies in the new US war on terror, and the remaining three are near-feudal monarchies with a growing sub-class of exceedingly disaffected youth, living on the declining royal patrimony derived from oil revenues. This arrangement does not exactly give confidence for security of supply.

Evidently the world is not about to run out of oil, having slightly more left than it has used during the first half of the oil age. Vested interests like to comfort us by pointing out that current reserves could support current production for more than 30 years. But a moment's reflection shows how absurd it is to imagine that production can be held constant for a given number of years and then stop dead, when production from all oilfields is observed to decline towards exhaustion. This leads to the issue of depletion, as we realize that discovery in any given area both begins and ends, reaching a peak in between. Production has to mirror earlier discovery in some manner. The natural pattern is confirmed by well-documented examples. Discovery in the 48 lower US states peaked in 1930, which led to a corresponding peak in production 40 years later. Discovery in the North Sea peaked in 1973, but advances in technology reduced the time-lag before peak production to just 28 years. Production in the UK is already falling fast, and Norway is not far behind. At the present rate of depletion, North Sea production will have halved within about ten years, forcing Europe to compete with the US, Japan, China, India and other importers for growing amounts of imported oil. Worldwide discovery peaked in 1964, and because this is fact and not supposition, it should surprise no one that the corresponding peak of production is now imminent.

Had the international companies retained control of the Middle East, with its abundant reserves of cheap and easy oil, they would have depleted it before moving on to more difficult and costly heavy oil, deep offshore resources, and oil from the polar regions. Such a natural progression would have alerted us to gradually rising cost and growing scarcity. But when the Western majors lost control of cheap oil resources through nationalization by producer countries, they turned to the above difficult and costly sources, and developed these in the shortest possible time. This left Middle East governments with the difficult task of managing "swing production" (the balancing of supply/demand gaps), in order to make up the difference between world demand and what the other, mostly non-OPEC countries could produce.

This in turn introduces the issue of spare capacity, which is another confused subject. No company or government has an incentive to drill oil wells only to shut them down or choke back production, but it is only such wells that can be restored at will. Capacity can be added in other ways, but that takes ever more work, investment and, above all, time. It now appears that the world had about run out of spare capacity three years ago, causing oil prices to begin to soar. They could have gone much higher had recession not intervened, cutting demand and reducing pressure on price. This pattern is likely repeat itself in the coming years, with economic recovery stimulating new oil demand until it again hits the falling ceiling of supply capacity in a vicious circle, re-imposing successive and deepening recessions. Figure III.9.1 shows the production of oil and gas from all sources under such a scenario.

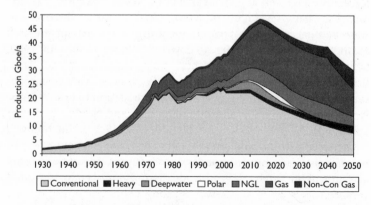

Figure III.9.1   Production of All Hydrocarbons Best-Case Scenario 2002.

It appears that the American government has yet to grasp these ineluctable facts, as it forges a new short-term foreign policy to meet domestic political pressures for cheap gasoline, considered by many voters to be as much a birthright as the right to carry a gun. The country now imports almost 60 per cent of its needs, which can only rise as domestic production continues its long, natural decline, save for a brief respite as its new deepwater oil begins to flow.

Future historians may look back and identify a degree of choreography in the present war on terror. The incident of September 11 was managed with such military precision as to shame a five-star general. The brief anthrax scare that followed it seems like an immaculately planned move to secure a popular mandate for military action. At all events, the US is paying for troops to defend a Colombian export pipeline; it was implicated in a failed coup to depose the Venezuelan president, who was taking a tough line on oil; it has overthrown the government of Afghanistan on a proposed pipeline route; it has established military bases around the Caspian oilfields; and it has invaded Iraq, one of the last places left with substantial oil reserves.

The US government may also be trying to hold down the price of oil by trying to undermine the confidence of the Middle East and OPEC in exercising their swing control. The otherwise respectable US Geological Survey offered support by issuing a report on the eve of a critical OPEC meeting, implying that on average 25Gb would be found each year between 1995 and 2025, with as much again coming from "reserve growth" from existing fields. If so, supply would more than meet demand, so that the producers would be forced almost to give the stuff away. The IEA and several governments were duped by this apparently scientific study, but the sad truth is that average annual discovery since 1995 has been only 10Gb, far short of the 25 claimed. The results are doubly damning, because the early years should be above average as the larger fields are usually found first. In fact, the balance between world consumption and discovery has been in growing deficit since 1981, as we eat into our inheritance from past discovery with a growing appetite.

The Stone Age did not end because we ran out of stone. It ended because we found better options as we moved through the Bronze and Iron Ages to reach the Computer Age at the pinnacle of the Industrial Revolution. This last chapter, which opened only 250 years ago, was fueled by cheap energy from coal, oil and gas. Now we have to retrace our steps to find ways to live with less energy from these sources. This time there is no better substitute fuel in sight that comes close to matching oil in terms of cost and utility. It is a shattering discontinuity, as option gives way to raw necessity. It is also a time of growing international tension, the first salvos of which are already being fired.

10
Oh Kyoto!

Andrew McKillop

If we ask when and how the Kyoto Treaty process began, several answers are possible: certainly at the 1992 Rio Earth Summit, perhaps in the 1980s or even 1970s, or perhaps only when the almost unlimited implications of accelerating climate change began to be grasped, later in the 1990s. The subject matter as well as the objectives of this strange and even incomprehensible Treaty have mushroomed into a spiral of political, strategic, economic, environmental, energy and North/South claims and counterclaims, with issues ranging from ice-cap melting, through energy use by national armed forces, to organic gardening and aquaculture, and the struggle of original peoples or anti-globalizers. Perhaps more importantly, despite ratification by about 180 countries, its *application* will probably be limited only to the EU countries, Japan, Canada and a few others.

FAR AWAY – IN THE REAL WORLD

As demonstrated by the World Trade Center attacks and US missile attacks on Baghdad, there is little limit to human aggression and violence, but few compare this violence to mankind's attack on, and destabilization of, the earth's climate through the alteration of its chemical makeup. The concentrations of carbon dioxide and other greenhouse gases have risen dramatically since Fossil Energy civilization began, extracting and burning carbon-based materials from the subsoil and accelerating the destruction of forests in its need to extract ever-more food for ever-growing urbanized populations. Change of the chemical constitution of the atmosphere has taken place on a planetary and geological scale: from about 270ppm (parts per million) in the early nineteenth century, the earth's atmosphere is today about 375ppm carbon dioxide. The last time such concentrations of CO_2 existed in the earth's atmosphere was at least 400,000 years ago, perhaps millions of years ago (see Part I).

We also do not know whether, although it is unlikely, such rapid build-ups ever occurred before in the earth's history. What is certain is that rising world temperatures and sea levels are not good news for mankind, as witnessed by increasingly disastrous cyclones, storms and floods, for example in Orissa, India and western France in 1999, in large areas of South and East Asia through 2000 and 2001, and in Central Europe and south and east Asia in 2002. In 2003 the records for high temperatures – monthly and seasonal – went on being broken in many parts of the world, with inevitable impacts on food grain harvests.

The detailed figures on how much temperatures will increase, and how rapidly sea levels will rise through ice-cap melt, and increased humidity above about 45° latitude, are the subjects of an energetic war of words, theories and models, but the UN's International Panel on Climate Change predicts a temperature rise of between 5°C and 8°C this century, with around a one-meter rise in sea levels. These background numbers drive many scenarios for loss of agricultural and urban land, regional climate change (including local cooling) and a certain but hard to predict reduction in agricultural productivity and food output. These potential, even probable, and *negative* changes all underlie the ultimate purpose of the Kyoto process, which is to have signatory nations reduce their current fossil fuel burn, returning to their fuel use levels of the early 1990s.

The Kyoto Treaty and its spiraling documentation – well over two-dozen kilograms and growing – ignores at least three facts that flatly contradict any fond hopes on its workability. Firstly, it has been virulently attacked by some leaderships (notably the Bush administration), but also readily accepted by most energy-intensive, post-industrial developed countries of the North because the process of reducing fossil energy burn will, they hope, shave oil and energy prices, giving their economies an additional lifespan before their inevitable and total restructuring. Secondly, nowhere does the Kyoto documentation acknowledge fossil energy depletion or its impacts. The earnest accent placed on "Clean Development Mechanisms" and on non-hydroelectric renewable energy development is surely laudable, but neatly avoids acknowledging that there will soon be no other non-nuclear alternatives to oil and gas (except small amounts of coal), and that energy use per person will inevitably fall dramatically within 35 years, particularly in the North. Thirdly, the vast majority of mankind is caught up in the conventional economic development based neoliberal political economy.

The potential for that economic, social and political trajectory to simply be replaced by a renewable-energy-based model for civilization is more in the realm of science-fiction or romantic folly than reality – yet there is no alternative.

When it is asked how this could have come about, how anthropogenic climate change could have occurred, there is usually silence and incomprehension. One certain reason, however, is that individually we can be unaware, or "choose to not know," much like other forms of life on this planet – for example, frogs.

FROG WARMING

Wim Wenger writes:

> The trick is to put the frogs live into lukewarm water, then gradually heat it. At any time until they lose consciousness, the frogs could, if alarmed, hop out of the pot and escape. But they don't, because they don't notice that the gradually warming water is heating up. Thank goodness we clever human beings, on this gradually warming Earth of ours, aren't stupid like those frogs! Well, as a matter of fact, we have noticed it – some of us, at least. We even have the data projecting the continuing ocean-rise swamping most of our major cities, and the pending disruption and collapse of our agriculture. Far more is going on than perhaps would suffice to alert a frog, even a very stupid frog. Even worse, at any given point we could fairly readily stabilize or reverse the temptrend, by any of a great many different means, but chances are that we won't, so settled are we within the walls of our stew pan.
>
> CommUnity of Minds, Energyresources, Yahoo!, June 24, 2002.

INFORMATION OVERLOAD AND DOCTRINAL REJECTION

Any Internet search engine should readily yield thousands of hits for "Kyoto and Global Warming." Likewise, if you want to explore the exotic technological fantasies of "carbon sequestration," or even "emission source and removal link coupling" or "the Berlin Mandate" – everything leads back to Kyoto. At inception, the concept was clear: we should first limit the increase rate of emissions, then progressively reduce the absolute quantity of CO_2 and other greenhouse gas emissions (notably methane, nitrous oxide, and other nitrogen–oxygen gases contributed by agro-industry). The concept of sequestration was added early on because the economic shock of suddenly reducing fossil fuel burning would probably kill the growth economy. "Sequestration" means the removal of greenhouse gases by trapping and storing them. But the ingenious and eccentric proposals made for sequestering greenhouse gases are primarily money-spinning ideas.

A sure reason for Kyoto and climate change dropping from the news headlines is that the whole subject area has far too much of a doom-laden ring to it. Implying that things aren't hunky-dory in our civilization and economy is considered to be in the poorest last, and denying any connection between fossil energy consumption and climate change has itself become a new money-spinner. What could be called "climatic revisionism" has become an important business, with tens of thousands of websites and a host of earnest journalists and commentators now making a living from climate change denial. Here is a typical example:

> The most reliable measurements show no change whatsoever in global temperatures over the last 20 years. What has changed is the perception that Global Warming makes a better scare than the Coming Ice Age. A good environmental scare needs two ingredients. The first is impending catastrophe. The second is a suitable culprit to blame. In the second case, the ice age fails and global warming is gloriously successful. It is not the destruction itself of Sodom and Gomorrah that makes the story so appealing but the fact that they were destroyed because they were so sinful.
>
> This is not a coordinated conspiracy but a fashion and a trend in which self-interest and ideology combine, and Green activists, politicians and journalists help each other to get more funding, more sensational stories and more enemies to blame. The climate of our planet is far too important for this nonsense. What we need is more genuine scientific research so that we can understand it better. If we do decide on the "precautionary principle" of keeping carbon dioxide levels stable, we can turn to those many technologies, proven or in prospect, which release no or little carbon dioxide. Nuclear power is the obvious first choice. There is no reason why the world economy cannot continue to grow and prosper ... But, for heaven's sake, let's start by telling the whole truth and giving all of the facts.
>
> Andrew Kenny, *Spectator*, London June 24, 2002.

THE CRITICAL MASS

When we consider the institutional response to global warming, it is very important to recognize that a kind of "critical mass" has already been reached: over 180 nations have ratified the Kyoto Treaty. The range of possible measures, some starting by 2008 and others in the period 2010–12, have ramified to implicate *every aspect of the economy and society*, in both North and South. The key term here is "implicate." Concerning on-the-ground changes to power generation methods or car mileage standards, for example, only a small number of countries will be affected by application of the Treaty's provisions. The Kyoto process has, however, gained political credibility – exactly because it is so pliable and vague that it can be used for a host of different ends, and offers a final and excellent excuse for decoupling oil consumption from economic growth, when or if necessary. When they wish to, political leaders can provoke a self-induced plunge into recession through interest rate hikes, but this time they could say they had to do it for the sake of the world's climate. Press communiqués and speeches at that time will surely announce, "This was vital if we were to comply with solemn undertakings we made on emission reduction." Politicians directing this willful nosedive into recession will need extra-hard-working justification for their decision; their real purpose being to limit oil price rises and curb inflation.

SOLEMN ENGAGEMENTS – FOR SOME

The Treaty sets out "solemn engagements" for emissions reduction. As Article 3 of the Kyoto Protocol to the UN Framework Convention on Climate Change puts it:

> The Parties ... shall, individually or jointly, ensure that their aggregate ... carbon dioxide equivalent emissions of the greenhouse gases ... [list] do not exceed their assigned amounts, calculated pursuant to their quantified emission limitation and reduction commitments ... [list] and in accordance with the provisions of this Article, with a view to *reducing their overall emissions of such gases by at least 5 per cent below 1990 levels in the commitment period 2008 to 2012.*
> UN Treaty N° 003912, May 2002. (Emphasis added; lists omitted.)

In other words, those countries which ratify the Kyoto Treaty (about 180) *and* which are engaged to reduce emissions (about 30) will start seriously limiting their greenhouse gas emission by 2008 (or in some cases a little later, in 2010). For some countries this will effectively be impossible – Canada and Ireland, for example, would need to make a truly heroic 25 to 30 per cent reduction of their fossil fuel burn or purchase "permits to pollute" from low emitters, or apply sequestration techniques (few of which work), or utilize other unspecified procedures by 2010–12. Alternatively, high emitters like Canada could demand extra time, renegotiate, change their accounting base for calculating emissions, or utilize other loopholes found throughout the 26kg of Treaty documentation.

There is a lot that can be tweaked before coming to the bottom line, where countries are invited to sign on for application of its provisions. There are so many possible interpretations that, after wading through the many thousands of pages of official and binding documentations, almost anyone, regardless of political leavings, would throw up their hands in exasperation and say: "Oh, Kyoto!"

WHAT PROPOSALS ARE ON THE TABLE?

Haggling has continued since 1997 on what could or should be set as targets for emissions reduction, depending on the country or group of countries. A list of proposals dating from the initial negotiations is given in the box.

PROPOSALS BY VARIOUS COUNTRY GROUPS TO LIMIT GHG EMISSIONS

Clinton Proposal: Developed countries would be obligated to reduce greenhouse gas emissions to 1990 levels between 2008 and 2012 (and a reduction of 34 per cent from projected levels in 2012), with further reductions afterward; unspecified participation would be required of developing countries; and an international system for trading permits for greenhouse gas emissions would be developed.

Japanese Proposal: Developed countries would be under legal obligations to reduce greenhouse gas emissions to 5 per cent below 1990 levels by 2012.

European Union Proposal: Developed countries would be under legal obligations to reduce emissions to 15 per cent below 1990 levels by 2010; such enforcement mechanisms as trade sanctions would be imposed on countries that fail to comply; there would be no change in developing country obligations.

Small Island States Proposal: Developed countries would be obligated to reduce greenhouse gas emissions to 20 per cent below 1990 levels by 2010.

Group of 77 Proposal: Developed countries would be under legal obligations to reduce emissions to 35 per cent below 1990 levels by 2020.

Source: "Greenhouse Gas Emissions," Heritage Foundation, US (1997–2002).

Proposals as of late 1997 – subsequently modified, with US refusal of ratification in March 2001 and "tentative" consideration of re-entry to the Treaty from May 2002.

Perhaps through an attempt to protect their creation from any accusations of partiality or exclusivity, but never stating that the major energy objective might *reduce oil and gas prices*, the teams drawing up this rambling Treaty included a multitude of incoherent, mutually contradictory ways and means. The aims are effectively a pick-and-mix range of shifting and imprecise goals for emissions reduction. While fossil energy burning stands out as among the greenhouse-gas-emitting activities identified in various Articles of the Treaty, these also extend to forestry, agriculture, animal husbandry, refrigeration and air-conditioning, fisheries and aquaculture, urban development and transportation, and many others. Being a post-neoliberal Treaty, it was de rigueur that there would be something in it for excited traders on world bourses, and this is more than amply provided by the Treaty's inclusion of *tradable licenses to pollute*.

TRADABLE LICENSES TO POLLUTE – OR TO PRINT MONEY

In very simple terms, if you are a country that in 2010 emits less greenhouse gases than allotted by Kyoto, then you have credits, termed "Hot Air Credits"; if you emit more than your quota or "baseline reference case" of greenhouse gases, then penalties will be levied, based on your percentage overrun from the base. For example "1990 + 9 per cent" is shorthand for saying that in 2010 your country exceeds its negotiated target level for greenhouse gas emissions by 9 per cent. Penalties to be paid (or licenses to pollute to be purchased) would be calculated from that base. These "tradable licenses to pollute" are the one alternative to paying penalties, and can be purchased as *credits* accruing to those lucky, well-managed, but in fact generally *poor* countries that are below their "baseline reference case" in terms of emissions. The purchaser (which can be a person or company, as well as a country) can then pollute in tranquility, happily knowing that the low-emitter country receiving payments can use this cash to invest in industrial equipment, consumer goods and machines, thereby *increasing* its own emissions. When or if the "low emitter" country gets above its own "baseline reference," it too will have to buy tradable licenses to pollute. We can note the tendency for this "market mechanism" to *increase* total emissions, but the Treaty's architects have anticipated that outcome.

The immense documentation contains various provisions that seek to limit this "perverse increase in total emissions," again by strange financial and fiscal measures, the details of which could keep an army of lawyers and a barrage of accountants quite busy and well fed for a number of years.

WHERE THERE IS CASH THERE IS AMERICA

Perhaps more strange is the US, which in March 2001 loudly slammed the door (as press and media would have it) on any idea that Uncle Sam would adhere to and ratify this socialist-minded attempt at hobbling America's energy appetite. From the very start of the Kyoto process, in 1992–97, the US has taken a great interest in the economic, fiscal and trading aspects of the Treaty. This feigned interest could be compared with UK government, media and public interest in the European single currency – so hateful to abide and so interesting to calculate exactly *when* the UK should dump sterling and rush to join the euro, as it unerringly will. The US Department of Energy (DoE) has played a lead role in analyzing what ratifying Kyoto might do for the US, especially where this concerns tradable licenses to pollute. This laudable attempt to find out if Uncle Sam might *profit* from joining Kyoto is hampered by the Treaty's own attempts at being inclusive and definitive; almost any Article will set targets, then bypass these with myriad escape clauses, which sometimes cancel each other out, but also sometimes reinforce each other.

One attempt at hard-edged definition is to set "cases" for increase or decrease in emissions, for any Treaty member, using the 1990 base of carbon dioxide and other greenhouse gases as the reference. This is used to develop scenarios for the necessarily vague quantities of tradable pollution that will be generated, thus creating possible values for the licenses or permits that might be shifted around world bourses at some stage in the future. The US DoE struggles to make dollars-and-cents sense out of the Treaty fog, but a typical extract from its analysis reads like this:

> The process of auctioning emissions permits would raise large sums of money. If permits were purchased from other countries, as is assumed in both the 1990+9 per cent and 1990+24 per cent cases, there would actually be two revenue flows – domestic and international. The carbon permit revenues remaining within US borders for each case are calculated as the

carbon permit price for that case times the level of carbon emissions in the 1990–3 per cent case. Thus, the number of carbon permits purchased domestically remains constant; only the price at which they are available varies across cases. Permits are assumed to be purchased abroad in order for US carbon emissions to continue above the 1990–3 per cent level. Therefore, the international revenue flow should equal the difference between actual emissions in the 1990+9 per cent (or 1990+24 per cent) case and those in the 1990–3 per cent case, times the carbon permit price in the 1990+9 per cent (or 1990+24 per cent) case.

US DOE Internet Home Page, June 2002.

Trading pollution credits might appeal to the business minded because of how this could operate. Investing in low-emission industries and energy production in a creditor country (called "Clean Development Mechanisms"), and then importing products or energy from that country in order to claw back credits, results in hostility. Business milieux in the US are clearly hostile to the Treaty, despite its being ratified by some 180 countries as of June 2002. American corporate analysts have seen and not liked what they imagine is the bottom line. As the most strident broadcast from thousands of Kyoto-hostile websites, every possible facet of the economy would be negatively impacted, according to them. A typical call to resist this socialist meddling with free enterprise can be found in this extract below, courtesy of the Heritage Foundation and WEFA, Inc. websites:

In a study for the US Department of Energy, the Argonne National Laboratory found that, if the climate change treaty were adopted, all US aluminum smelters and paper producers would be forced out of business; 30 per cent of the basic chemical, steel, and cement industries would move to developing countries or be forced to close; and petroleum refinery output would be reduced by 20 per cent within 20 years. [The Clinton administration's] proposal to address global warming would result in lower economic growth in every state [of the US] and nearly every sector of the economy. This lower economic growth would lead to reduced employment and deteriorating wages. Before committing the United States to such an austere economic course of action, Members of Congress should examine the relevant studies closely and assure themselves that the benefits of adopting the global warming treaty would be worth the inevitable costs of curbing greenhouse gas emissions.

BIG BUSINESS AND THE MILITARY DO NOT LIKE KYOTO

Estimates by various US think-tanks and Republican-oriented consulting groups have heaped on the bad news, arriving at figures of "3,300 billion dollars of lost national output through to 2020" if the US were to ratify and apply the Treaty from 2010. Perhaps worst of all, some shocked defenders of the US's integrity and world role (that is, its war capacity) discovered, mainly by imagination, that ratifying this Treaty would *severely* hamper national security:

> The Pentagon estimates that a 10 per cent cut in its fuel use, to reduce carbon dioxide emissions, would reduce tank training by 328,000 miles per year, flight training and flying exercises by 210,000 flying hours, and the number of steaming days – days on board ship in port and at sea for training and naval exercises – by 2,000. These reductions would substantially hamper military readiness – adding as much as six weeks to the time air forces and tank corps need to deploy in a time of crisis. What would our enemies be doing while our troops got up to speed? And a 10 per cent emission cut would be only one-third of the military's share of the cuts needed to meet our commitments under the treaty.

Before giving these rather precise figures on reduced tank mileage, the author of these shocking figures, H. Sterlin Burnett ("Global Warming Treaty Threatens National Security," *Investors Business Daily*, October 15, 1998) indicates the exact fossil energy dependence of the US military:

> What does a treaty ... to prevent human-caused global warming have to do with the US military? More than you think. It turns out that the federal government is the United States' largest consumer of energy. And 73 per cent of the federal government's energy use goes to the Defense Department. Finally, if the Pentagon [was able] to get a blanket exemption, that just means the private sector will have to make even deeper cuts to make up for it. Harming the US economy would not seem to be any more in our interests than hog-tying the US military in case of a security threat.

To be sure, the military implications of Peak Oil are immense. Declining supply of cheap energy will accelerate changes in the *types* of wars human beings fight. The impacts of Peak Oil on "classical" or "conventional" warfare are probably more important to *how the world changes* over the next 25 or 30 years than the boom-and-bust that energy depletion promises for the shaky foundations of current "prosperity" in the high-energy nations of consumer civilization. Modern warfare is next to impossible without gasoline, jet fuel, electricity and energy-intensive industrial products. Apart from inevitably increasing urban warfare, the military security of any nation – except those that will sink beneath the waves due to sea level rise (perhaps a dozen by 2100) – will increasingly mean dependence upon oil and gas producing regions or countries. These countries will be not only economic prizes, but also strategic goals that must be held to ensure future war readiness.

THE FAR-OUT WORLD OF CARBON SEQUESTRATION

It is hard to overemphasize the spreading of the "Kyoto message" into our lives, rather like the ever-rising mass of carbon dioxide in the earth's atmosphere. We are now invited to hunt down the element carbon, as it is emitted by everything from the family car to the family pet, from uranium mining to supply nuclear reactors to throwaway gas cartridges for picnic barbecues. Consequently, a new branch of thinking has emerged, named "carbon sequestration." This rogue carbon must be sequestered – that is, taken out of and away from the atmosphere. This fantastic task has already generated projects showing the almost unlimited human capacity to imagine that wishes can be fulfilled – and if not, one can at least turn a profit.

It should therefore not be surprising to find, for example, that the international federation controlling world Formula One motor racing (FIA) purchased 5,000 tons of carbon dioxide in 2001 to be sequestered and guarded by "an indigenous people organization in southeastern Mexico," these five kilotons being estimated as the quantity of CO_2 that the 2001 season of F1 race events would generate.[1] Other projects, almost as wacky, abound in various parts of the world; for example the "Oporto–Rotterdam Sequestration Project" by which investors can "help" pay for tree planting, to absorb carbon dioxide, thus "sequestering" it, starting with a former football field in Oporto. Those participating in this project, by sending at least 10 euros, will receive payment receipts in the form of "bills" issued by the Holland-based ING Bank.[2]

The receipts are described as tradable – that is "able to be sold to other business or private users with an obligation to reduce their emissions."[3] This activity, if not very practical in sequestering carbon dioxide through fixing it in the form of wood (which ideally should never be burned or allowed to rot and thus should be kept away from the biosphere when dead) can at least give a rosy feeling of doing things for the environment and climate, and might turn a profit, to the greater glory of the financial players. The same website inviting purchase of 10-euro carbon sequestration bills indicates that EU per capita average carbon dioxide emission in 2001 was 666kg per month, or about 8,000kg per year, while also indicating the number of forested football fields, whether around Oporto or elsewhere, that would be needed to sequester all CO_2 emitted by the 600,000 inhabitants of Rotterdam. The total is 995,215 standard-size football fields forested *each year*, each planted with a mix of fruit, conifer and deciduous trees, yielding about 65 million hectares of woodland, each hectare absorbing about 70 tons of CO_2 per year.

CARBON SEQUESTRATION AND CARBON CONSUMER CIVILIZATION

If you think we are already into big figures just for Rotterdam, then you are right, because this forest area is about 12.5 times *the total land area of the Netherlands* – some 41,700 square kilometers. Going a little further, we can compare annual world release of carbon dioxide from fossil fuel burning, at about 29 billion tons, with total world biomass production, which is between 25 and 33 billion tons per year. If global CO_2 emissions were to be "sequestered" as sugar cane, which is one of the most productive methods for carbon fixation per unit area, then the annual sugar cane pile would form a green pyramid 35km wide by 35km long, and 3.5km high. If global CO_2 emissions were to be captured and prevented from entering the atmosphere through the "forest capture" route, then somewhat more than the entire land area of Europe (4.3 million square km) would have to be forested *each and every year*, and the wood never burned, or, at least never burned and never allowed to rot. This could be done, for example, by the carbon sequestration technology of storing it in sealed caverns or sunk in sealed bundles underwater, or by yet other sequestration "technology" or procedures. To be fair, even the most ardent fans of carbon sequestration do not advocate total elimination of carbon dioxide from every stack, exhaust and vent emitting combustion products from fossil fuel burning, focusing on single, large sources of emission such as power plants and cement factories. The Norwegians are now pioneers of sequestration technology, and in late 2003 had demonstrated equipment and methods in the North Sea for reducing emissions from oil rigs (of flared gas) by injecting it into empty oil and gas strata, for an amount of carbon equal to about *0.1 per cent of total North Sea oil and gas related emissions*!

All this is purely symbolic, however well-intentioned. Nevertheless, research and development continues into carbon sequestration, while the culprit runs free in the atmosphere for at least 100 years before being captured by the earth's carbon fixation and cycling systems – first in the oceans, and taken by biological and geological capture.

THE BIG PICTURE

All biological activity is based on carbon uptake or cycling of carbon-based chemical compounds, and the emergence of life on this planet from a static and wholly inorganic environment changed the atmosphere's composition, as well as that of the seas and the top layers of earth's crust, by creating soil. Annual biomass production of all land surfaces runs at about 25–33 billion tons of fixed carbon per year, with ocean biomass at about 35–45 billion tons per year, with the "reservoir" capacities of existing carbon in the land and ocean areas of the planet being around 10^{15} tons. Perhaps more interesting is that at least 100 billion tons of carbon dioxide from the atmosphere dissolves in the oceans each year, and is replaced by about the same quantity of CO_2 from the oceans. The constant addition of CO_2 by fossil fuel burning – and also through erosion, forest conversion to croplands and other human activities releasing or generating carbon dioxide – is without question increasing biomass production on land and in the seas. Even there, however, we have to qualify this. It has been proved that tree growth rates across Europe have increased by the stupendous rate of about 50 per cent in 100 years,[4] and carbon dioxide enrichment of the atmosphere, as well as global warming, is certainly a cause of this change. However, while erosion (due to intensive, single-crop, energy- and chemical-based farming, urbanization and road construction) increases CO_2 emissions, it also *reduces* fertile land areas able to support vegetal life. World annual soil erosion has now reached the stupendous figure of about 30 billion tons per year, although D. Pimental et al. give estimates for worldwide soil loss as approaching 75 billion tons per year.[5] In some ways, then, this is a race between destruction of the thin, fertile and carbon-rich part of the earth's crust we call soil, and increased biomass production and productivity (intensity of growth per unit area) due to "doping" by carbon dioxide. As noted previously, world fossil fuel burning produces about 82 million tons of carbon dioxide *per day*, and the annual amount produced, about 29 billion tons, is very close to *annual planetary biomass production*.

The only winner in this race to erode massive quantities of the earth's surface while "enriching" the atmosphere with CO_2 and "doping" (or over saturating with CO_2) plant growth is ocean phytoplankton, and then only in areas not seriously affected by pollution from industrial and urban effluents. Land areas of the planet are almost certainly net losers in plant biomass production, due to man's presence. However, in the world's seas and oceans the losers include fish stocks, where overfishing – made possible only by the intensive use of fossil fuels – has reached critical limits. These factors in the equations deciding what human population sizes, densities and lifestyles can and will survive the final energy crisis are discussed elsewhere in this book by Goldsmith (Part I), McCluney (Part III) and Newman and Trainer (Part V). What counts in the changes predicated by the Kyoto process is *unrelated* to the accelerated atmospheric and climatic changes now occurring. These changes were the ultimate result of fossil fuel burning from the 1860s up to now, shortly before Peak Oil. In theory, there is no need to consume the other half of the world's fossil fuel endowment in the next 35 years, but this outcome is the most likely, and the annual decline in physical production beyond Peak Oil, this decade, will make economic and geopolitical crisis a near certain bet. The life expectancy of the Kyoto Treaty is probably much shorter than that.

11
Renewable Energy Limits
Ross McCluney

The more we get out of the world the less we leave, and in the long run we shall have to pay our debts at a time that may be very inconvenient for our own survival.

Norbert Wiener

The world faces serious energy and environmental problems. These crises are inextricably linked. The processes of extracting, processing, and burning fossil fuels generate copious pollution of air, water, and land. Fossil-fuel-derived energy is at the heart of other environmental problems. Fossil energy powers the bulldozers clearing rainforests; it runs the tractors and other farm equipment of industrial agriculture, compacting and mineralizing soils, thus increasing their susceptibility to erosion; it provides the fertilizers and pesticides used by intensive food-production systems, "freeing" people for other pursuits. It enables the spread of cities dependent on motor transportation, and their attendant environmental impacts. The depletion of our fossil resources over the course of the twenty-first century is a serious concern requiring considerable attention from individuals, organizations, and governments.

Most leaders seem to think that science, technology, and some minor alterations in public policy will be sufficient to prevent these problems from adversely affecting humanity as a whole. Even solar enthusiasts seem to think that most of these problems will go away if everybody would just buy (or make their own) solar energy systems. Other people believe that the threats facing mankind are much more serious and need more urgent attention and deeper action. These people are concerned that leaders in all sectors of society are failing to see the long-term threats and refusing to do much about them, save for some recycling, a few pollution restrictions, a little energy conservation here, some solar energy there. Growing evidence is leading us to believe that the pessimistic view is the correct one, that current reforms will be insufficient for the long term.

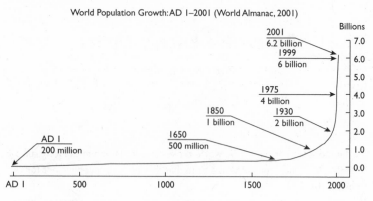

Figure III.11.1   Exponential growth of world population

ENERGY AND POPULATION

As illustrated in Figure III.11.1, world population expanded only gradually until the Industrial Revolution, reaching about 1 billion around 1850. Then came oil, and later gas, added to already established coal production, enabling the Industrial Revolution, the machinery of which was driven by abundant cheap energy from fossil fuels. In the span of about 3 million years of human life, the current spurt of fossil-fuel use and depletion has occupied only a tiny portion.

The recent exponentially rapid growth of world population tracks growth in the use of fossil fuels. Campbell described it thus:

> The abundance of energy has allowed the human population to expand greatly, multiplying three-fold during the lifetime of the present Queen of England. A new subspecies, called *Homo hydrocarbonum*, evolved ... [and] will certainly be extinct by the end of this century ... We are not about to run out of oil, but production is close to peak. The transition will represent an unparalleled discontinuity as the growth of the past gives way to decline in the future.[1]

Homo hydrocarbonum will be replaced by *Homo somethingelseonum*. What that something else will be is the subject of this chapter.

ENERGY AND FOOD

Ecologists tell us that food is an important controlling factor in population dynamics, including die-offs of non-human animal populations, as well as our own. Because of the high fossil-fuel subsidy to industrial food production, the need for fresh water for irrigation (mostly pumped using fossil fuels), and the increasing purification of water, necessitated by the pollution generated by such intensive agriculture and urbanization (also dependent on fossil fuels), it seems clear that human populations will track the availability of fossil fuels downwards, following Peak Oil.

In spite of this, our leaders seem to take heart in signs of increasing agricultural productivity in many parts of the world. Other areas, however – with lands not as suited to high-yield agriculture, or suffering shortages of fossil fuels and energy-based infrastructures – experience terrible famine and poverty. Further, much of the increased productivity has been accompanied by ever-increasing dependence on fossil resources. As these resources become short in supply, agricultural productivity can be expected to decline. Even before we reach the down-slope of fuel availability, food shortages for many hundreds of millions of people is the daily reality.

On October 15, 2002, the United Nations announced that "[p]rogress in reducing world hunger has virtually stopped, and mountain sources of fresh water essential to food production are melting away due to global warming." Additional dire statistics were presented in *The State of Food Insecurity in the World 2002*, the annual report of the UN Food and Agriculture Organisation (FAO). It stated: "As a result of hunger, millions of people, including six million children under the age of five, die each year." The FAO estimates that there were around 840 million undernourished people in 1998–2000. Of these, 799 million are in developing countries, 30 million in "transitional economy" countries, and 11 million in the industrialized countries.[2]

Each year, chronic hunger and malnutrition kills millions of people, stunts development, saps strength and cripples victims' immune systems. Where hunger is widespread, mortality rates for infants and children under five are high, and life expectancy is low. Even if agricultural productivity were increasing in all parts of the world, we would still have a serious problem. The reason is that high-yield species and energy-intensive methods of production are unsustainable without massive petroleum inputs. If the petroleum inputs go away, or even just stop growing and remain stable for a while, this will exert strong negative pressure on the world's food production systems, as we shall likely be witnessing in the next few years.

Writing in 1999, Pimental et al. pointed to declining available cropland:

> In 1960, when the world population numbered about three billion, approximately 0.5 hectare of cropland was available per capita worldwide. This half a hectare of cropland per capita is needed to provide a diverse, healthy, nutritious diet of plant and animal products – similar to the typical diet in the United States and Europe. The average per capita world cropland now is only 0.27 ha, or about half the amount needed according to industrial nation standards. This shortage of productive cropland is one underlying cause of the current worldwide food shortages and poverty.[3]

The ratio of petroleum energy to solar-derived food energy was explored for the US by Steinhart and Steinhart in 1974. They found that the total energy subsidy for all food types in the US was 1,000 per cent: every calorie of food energy consumed it took about ten calories of fossil-fuel subsidy to grow, harvest, process, and deliver.[4] Though it is not nearly this high in world average terms, the fossil-fuel subsidy required simply to maintain current high-yield food production is an integral part of world agriculture, and a warning of difficulties to come. Fossil-energy subsidies to food produced by various methods are shown in Figure III.11.2. For the most energy-intensive, developed countries, the solar energy content of the industrially produced food we eat is so low that, in reality, we are not "eating" solar energy but are "eating" oil.

As petroleum production starts declining, as water shortages grow, and if agricultural yields drop due to the environmental effects of fossil-fuel based agriculture, such as accelerated erosion and climate change, food output will decline and prices will rise. If world population continues its currently inexorable increase after Peak Oil, then per capita food supplies can be expected to shrink rapidly. This will pose threats of serious economic and social upheaval – as the number of starving people increases and as others seek violent means to correct the massive global imbalances in wealth and food security. Those with economic power probably won't suffer much. Though food will be more costly, they will still be able to buy it. The less well off, however, may not be content to starve while others are well fed. Wars of terrorism can be expected to grow, and be augmented by new wars between the haves and have-nots. The resource wars have already begun.[5]

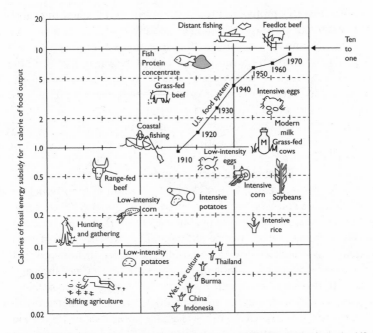

Figure III.11.2 The link between food production and fossil-fuel subsidy in the US. Abundant food makes the large human population possible. Food is abundant largely because inexpensive fossil fuels are abundant.

TO THE RENEWABLES

Many people hope we can avoid the threatened difficulties by switching from petroleum energy to solar energy, backed by increased energy conservation. The possibilities seem promising. New technologies include wind- and solar-powered electric generating stations, solar heating systems, ocean energy systems of several kinds, and possibly geothermal energy. The threats of future upheaval add urgency to work in the fields of energy conservation and solar energy production. Partly because of this urgency, research, experimentation, and use of energy-efficient and renewable-energy technology is very exciting, and is moving forward, though slowly. Important demonstration projects around the world demonstrate stimulating work, opportunities to create new low-energy-consuming systems, challenges to develop and install many solar technologies, and the potential for contributing to the betterment of humankind.

It was just this motivation that led me to leave a promising career at NASA to pursue research in solar energy 27 years ago. My solar work has been stimulating and rewarding, in spite of a diminution of US federal and worldwide governmental and intergovernmental funding for solar research over the years, relative to the inflation-adjusted levels of the 1970s. The expected expansion of research funding failed to materialize, and the field is not growing as rapidly as it should.

Perhaps for the same reasons, current use of solar energy technology is limited, compared with total energy use. As energy price increases result, sooner or later, from declining oil production, the use of both renewables and energy conservation can be expected to increase, perhaps enough to make up for declining oil production. Coal reserves could be used to extend our fossil-energy future somewhat, following the decline of oil, but switching "backwards" to coal – a less desirable source – will be difficult and unwise. We might try to expand our use of nuclear fission, but this entails further growth of nuclear waste disposal problems. Along with this comes increased susceptibility to nuclear terrorism, and to large, even catastrophic accidents due to the aging stock of nuclear reactors that must be decommissioned. Nuclear energy is a false hope, narrowing our real choices to conservation and solar energy.

TO THE FUTURE

Although energy conservation and solar energy are to many our great hope for the future, these technologies do have some limitations. As we seek to make solar energy and conservation popular, the number of people demanding more energy use is also increasing worldwide, exacerbating the problem. Developing countries generally seek to emulate the profligate energy consumption patterns of the industrial world. As an example, the US consumes more energy per capita than any other nation on earth. If current solar and conservation development efforts succeed, and if world population is allowed to continue increasing at its current annual rate of 1.4 per cent (with a doubling time of a mere 47 years), whatever gains we see through energy efficiency and renewable energy expansion threaten to be offset by increased demand for more energy worldwide. Energy demand increase results from a combination of population growth and growth in per capita energy use.

I was in Beijing's Tiananmen Square a few months before the June 1989 uprising there. My feelings were divided as I watched news reports of the event back home. On the one hand I was excited for my new Chinese friends that a more open or even democratic society might finally come to China. This was countered by a realization that, whatever happened on the political scene, this would be followed by rapid economic expansion and material prosperity – along the lines of the energy-wasteful and polluting industrialized world. "What would happen," I asked myself, "if even 20 per cent of China's 1 billion people suddenly purchased automobiles?" (see Chapter 16). In the 16 years since, China has modernized through accelerating its industrialization. So my fears of environmental threats are becoming real, in spite of some good efforts by the Chinese government to develop along a less polluting, less wasteful path.

The population of China is now 1.3 billion. That of India is 1.1 billion. Together they constitute over a third (37 per cent) of the 6.26 billion world total. The rapid industrialization of these two countries will place a heavy burden on the world's ecosystem. Just as we are attempting to improve energy use efficiency and make the switch (albeit slowly) to renewables, the demand for energy worldwide is growing as reported by the UK Atomic Energy Authority and the US EIA.[6] The first of these forecasts energy use patterns considerably into the future, so can only indicate wishful expectations.

ENERGY TRANSITIONS

There have been several energy transitions in the past. Previous transitions from wood to coal, and then to petroleum and natural gas, were fairly rapid and singular. No other previous form of fuel competed very well against each newly discovered one. Here are some reasons:

- New fuels were in most respects better than former fuels. They were more concentrated, easier to store and transport.
- All alternatives to the new fuel were more expensive, therefore less economically viable.
- There was essentially universal agreement that the new fuel was better, so there was little dissension over the transition to it.
- With the exception of coal, the new fuel was generally cleaner.

Hydropower and geothermal energy, although clean, obviously cannot be transported, except as the end-product electricity; and their availability for new energy generation is geographically limited.

We are finding that the next energy transition, away from the fossil fuels, will be towards not just one source but to *a variety of different ones*, all possessing only the fourth of the advantages cited earlier. The new transition will be more difficult than the previous ones.

DIRECT USE OF SOLAR ENERGY

Solar radiation has been proposed as the great new replacement for fossil fuels. Though relatively clean (except for some pollution during manufacture and disposal of the hardware), solar (and sky) radiation is a more diffuse, less concentrated resource than petroleum, and is also not easily or cost-effectively stored. To store it requires conversion to some other energy form, such as heat, electricity, or chemical energy. This is difficult and costly, and has inhibited the spread of solar technology; in some cases it even has serious environmental drawbacks.

In a partially "solarized" society, one can envision using our remaining, but declining non-renewable energy sources for storage – to fill in during the times when solar energy is not available, at least temporarily avoiding the need for solar storage. In a more fully solarized economy, electricity generated with solar-derived stored energy could provide the backup. Biogas and methanol from (solar-powered) crops come to mind for this use. The problem is that obtaining energy from these sources is very land-intensive – even more so than direct solar or wind energy production.

Hydrogen gas is a much touted means of energy storage. It can be solar-produced, through solar-powered electrolysis of water; it is clean-burning and non-toxic. However, it is the lightest of the chemical elements and difficult and costly to store and concentrate. Research is in process to find ways of storing and releasing hydrogen chemically, avoiding the need for expensive, heavy, and potentially dangerous high-pressure storage tanks. If these problems can be overcome, our hopes for hydrogen as a portable fuel may be realized, but it will be neither easy nor inexpensive. It is true that a copious quantity of energy arrives from the sun each day. It falls all over the earth, but harvesting it directly, in sizeable quantities, means that it will be diverted from alternate uses in nature. Massive use of solar energy will require alteration of vast areas of the land and water surfaces of the planet, changing biosphere systems in the process.

Solar radiation is not the "be all and end all" energy solution for the world. In addition to the dilute nature of solar energy itself solar conversion systems currently require fossil fuels to manufacture them. As Baron described it in 1981, "[a] major solar energy cost component is the cost of non-renewable resources of oil, natural gas, coal, and nuclear energy consumed in producing and constructing the systems for solar heating and solar electric plants."[7] The proper assessment of a proposed solar technology should include a determination of the system's net energy production. In 1978 Peter Knudson described this as follows: "Net energy analysis, in its broadest sense, attempts to compare the amount of usable energy output from a system with the total energy that the system draws from society."[8] I offer my own definition: it is the magnitude of the solar-energy-derived output minus the non-renewable energy drawn from the earth needed to make and operate the solar-energy system. There are calculations showing that the net energy of solar collection and distribution systems in some cases is negative. Also, at the end of their useful lives they have to be dismantled and recycled (with additional expenditures of energy).[9] If the net energy output of a solar technology is negative, logic leads us to the question, "Why bother?" In such a case, wouldn't it be better to use the fossil energy directly, rather than lock it up in an inadequately producing solar-energy system?

This is a controversial topic. Even if the calculations are correct for some situations, there can be value in storing present-day (less expensive) fossil energy in solar collection devices, as a hedge against future depletion. Continuing with this argument, since we are going to use up fossil fuels anyway, why not invest that energy in the manufacture of renewable energy systems, so they can go on producing power when the fossil fuels are depleted? Ultimately we would like to remove the non-renewable energy inputs from the manufacture of solar energy systems altogether. This leads in turn to the idea of a solar "breeding" system – using solar energy in the mining and processing of ore, and the manufacture of solar-energy systems, thereby reducing or eliminating the fossil-fuel subsidy. Solar-energy systems produced by such a system will be strong net energy gainers. For such a strategy to be successful, the solar-powered mining and manufacturing industry must be completed before the fossil fuel sources are gone (or become exorbitantly expensive). This might enable the establishment of a society based solely on solar energy, using solar energy alone to recycle worn out solar energy systems. Ultimately we might be able to develop a completely sustainable process not requiring the extraction of further fossil fuels from the earth to keep it going, as long as requirements for "fresh" inputs are continually reduced. Of course, a fundamental assumption of such proposals is that the earth's human population is reduced to a sustainable level, supportable by totally renewable energy systems alone.

SOLAR POLLUTION

The non-renewable energy subsidy required to make and operate solar-energy systems is not the only issue. The degree of environmental destruction associated with an energy-consuming or producing system of any kind is also critical. As Baron pointed out in 1981, "[e]ven more serious would be the impact upon public health and occupational safety if solar energy generates its own pollution when mining large quantities of energy resources and mineral ores."[10] Some solar energy manufacturing processes produce toxic or other waste products which have to be recycled, discarded, or otherwise rendered benign. Clearly, we'll have to pick and choose among the solar alternatives to find the least environmentally impacting ones, and work hard to improve all the rest.

SOLAR LIMITS

There are physical limits to the production of energy from direct solar radiation. At an absurd extreme, we clearly could not cover all available land with solar collectors. A more reasonable limit would be to fill existing and future rooftops with solar collectors. From data provided by the US EIA, I estimated the total combined commercial and residential building roof area in the US in the year 2000 at 18 billion m^2. From a National Renewable Energy Laboratory website, I found that the approximate annual average quantity of solar energy falling on a square meter of land area in the US is about 4.5kWh of energy per square meter per day. Multiplying this by 365 days in a year, then by the 18 billion m^2 roof area figure, yields the total energy received by rooftop solar systems in this scenario: 2.46×10^{13}kWh per year, or 84 quads per year. This is just a little below the 102 quads per year US primary energy consumption figure. Not all roof area is usable, however. Roofs sloped away from the sun's strongest radiation, shaded by trees and other buildings, having interfering equipment, or insufficiently strong to support solar equipment, cannot be used in this way.

The conversion from primary to end-use energy is not perfectly efficient in the case of either renewable or non-renewable energy sources. Both the 102 and the 84 quads of primary energy must therefore be reduced when converting them to actual end-use energy. It is difficult to determine accurate average conversion efficiencies for all technologies in both categories, but they are not likely to be widely different. Thus, the conclusion should remain valid that meeting total US energy needs with 100 per cent direct solar energy would require roughly *every single square foot* of roof area, of all commercial and residential buildings in the country.

Since most current roofs were neither designed nor built to carry the loads of (and wind loading on) solar collectors filling them, nor are they all exposed adequately to the sun, it is very unlikely that we could achieve the goal of a 100 per cent solar economy in this manner. For every acre of existing rooftops which cannot be filled with solar collectors, an equivalent acre would have to be found elsewhere; and renewable energy from other sources would be needed. If the US population continues to grow, pressure will continue mounting to expand developed land areas into what are currently agricultural and wilderness areas. If the plan is to convert as much as possible of the US energy economy to direct solar energy, solar collector farms will join in the competition for new land to be opened up for this use.

In order not to have to convert agricultural or natural habitat areas to areas for engineered solar production, one would have to find other, already developed areas for erecting these solar collectors, such as street and highway corridors and parking lots. While the shaded areas might be attractive to those having to drive and park in the hot sun, it is probably not economically feasible under current financing conditions. A number of other objections to this possibility can be expected, leading to pressure to convert agricultural and wilderness areas to solar production "farms."

WHAT ABOUT THE DESERTS?

Solar energy development needs for large land areas are a serious problem, if you wish to power the world with it. We have learned from other experiences about the adverse environmental and social impacts of growth and land development. Solar energy is unlikely to be exempt from all these impacts. A common reaction to the problem of finding areas for large solar energy collection systems in forests, on farms, or in developed areas is to point out the vast "unused" desert areas around the globe, suggesting that these would be good places for solar collectors. Some desert areas can certainly be used for renewable energy technology, but there are limits. Deserts are not devoid of wildlife; they contain many species of flora and fauna, adapted over millions of years to desert conditions. There is a limit to how much desert we can cover with solar collectors.

INDIRECT SOLAR: RENEWABLE ENERGY TECHNOLOGIES

In addition to the solar energy which can be collected directly, there are several indirect sources of this important resource. They include the *wind*, powered by differential heating of the earth's surface, *ocean currents* (produced by a similar mechanism), *hydroelectric* energy, produced by solar-powered water evaporation and condensation into rivers, *tidal currents*, *ocean thermal energy conversion* (based on solar-heated surface layers of the tropical oceans), and ocean *waves*, driven by the wind and carrying energy with them as they approach the shoreline.

Waves and thermal-driven currents offer a degree of natural solar concentration. Tidal currents are also concentrated in some locations. Solar-derived wind, pushing the sea over large distances, increases the height and energy content of waves. This energy can be extracted downwind, where the waves are most intense. Thermal currents can be focused between land masses, thereby concentrating the speed and energy content of the moving fluid. *Geothermal* energy is another possible source. It uses the heat from deep below the earth's surface to produce electricity. Let's take a look at each of these renewable technologies.

Wind Power has now become economically viable for areas experiencing adequate average wind speeds. Wind turbines are being erected on the land in many locations around the world. Due to the difficulty of finding onshore sites, and other factors, they are increasingly being sited offshore as well. The February 2002 issue of *Renewable Energy World* describes an example.

> Offshore wind farms promise to become an important source of energy in the near future; it is expected that within ten years, wind parks with a total capacity of thousands of megawatts will be installed in European seas – the equivalent of several large, traditional coal-fired or nuclear power stations. Plans are currently advancing for such wind parks in Swedish, Danish,

German, Dutch, Belgian, British, and Irish waters. Outside Europe there is serious interest in such developments on the US East Coast, and in Australia the resource off the Tasmanian coast seems to be attracting attention, also.[11]

Many proponents of wind-power claim that wind farms on land can co-habit with agriculture, and that leasing such land can be a valuable source of income for the farmer. A substantial number of wind turbines have been installed in windy desert areas, and more are expected. Wind is generally variable in its speed, however, so it is not the most suitable source for what is called "baseload" electricity generation – that non-variable core power component that forms the backbone of electric utility operations.

Ocean Currents, such as the Florida Current, the part of the Gulf Stream flowing northward past the Florida peninsula, carry enormous quantities of kinetic energy in their motion. There have been several proposals to develop this resource, to place ocean turbines in the strongest of currents and feed the energy generated onshore. According to Practical Ocean Energy Management Systems, Inc., "The first large ocean-system proposal is for a 2.4-mile system that would link Samar and Dalupiri islands in the Philippines. The Dalupiri project is now estimated to cost $2.8 billion, produce 2,200 megawatts at tidal peak and offset 6.5 million tons of carbon dioxide a year."[12]

A number of years ago I made a "back-of-the-envelope" calculation of the available energy in the Florida Current. The kinetic energy transported through a cross-sectional area by a fluid of known mass and density is the product of its kinetic energy per unit mass of moving fluid and the mass flow rate through that cross-sectional area. The energy flow rate, per unit area, is proportional to the *cube* of the current speed. All the kinetic energy contained in the flowing water cannot be usefully extracted, or the flow would cease. It should be possible to extract enough energy to slow the stream by about 50 per cent or so. In this case, approximately 88 per cent of the available kinetic energy would be extracted. I estimated the electrical generation potential of the Florida Current, between Miami and the Bahamas. The Gulf Stream flows northward through the straits at a speed ranging from 2 knots at the edges to over 4 knots in the middle of a 20-nautical-mile (37km) width off Miami. This yields an approximate average kinetic energy per unit of cross-sectional area transported by this current in the order of 2,000W/m². If we assume an 88 per cent conversion efficiency (slowing the current by a factor of two in velocity), and that we extract this energy from the surface down to a depth of ten meters over the 37km width of the current (370,000 m^2), then the total electrical power output for a 100 per cent efficient electricity generator would be in the order of $0.88 \times 2,000 \times 370,000 = 651MW$. If we choose a 30-meter depth, the power generation would be three times larger, or 1.9GW, a large electrical generation capacity.

According to Practical Ocean Energy Management Systems, Inc.,[13] ocean currents are one of the largest untapped renewable energy resources on the planet. Preliminary surveys show a global potential of over 450GW, representing a market of more than US$550 billion. The *Proceedings of the MacArthur Workshop on the Feasibility of Extracting Useable Energy from the Florida Current*[14] can be consulted for more information about the resource potential of the Gulf Stream. The US DOE in 1979 funded a study of the Florida Current energy potential that used a much larger cross-sectional area across the Straits of Florida, involving 132 turbines with duct exit areas of 22,900 m^2, for a total cross-sectional area exceeding 3 million m^2, and having a maximum rated electrical output of 10,000MW from the 2.3m/s current. The study estimated that in the presence of variations of current strength, the total effective power output would range from 2,000MW to 6,000MW, equal to the output of several large conventional power stations. It is doubtful, however, that the economic value of the electricity that might be generated, though large, would be sufficient to offset the huge costs of construction, including anchoring underwater structures in strong current in deep water, and dealing with whatever environmental consequences might be produced (including potential impacts from diverting the current's warm water away from northwest Europe, with unpredictable climatic effects).

Tidal Energy can be extracted by placing turbines or other current energy extractors in or across the mouths of estuaries experiencing large tidal excursions. Energy in the flow of ocean water in and out of the estuary can be extracted and turned into electricity. A working power plant of this type is located in France. It produces 240MW of power via a "barrage" across the estuary of the river Rance, near St Malo in Brittany. The plant went online in 1966, and supplies about 90 per cent of Brittany's electricity. This is a fairly unique installation; it is doubtful that it could be duplicated at reasonable cost in many places around the world.

For tidal differences to be harnessed into electricity, the difference between high and low tides should be at least 5 meters, or more than 16ft. There are only about 40 sites on the planet with tidal ranges of this magnitude. Tides of lesser magnitude, however, could be used to produce usable power. Turbines placed under the water, grounded on the bottom, could allow shipping to pass overhead while still generating power. Currently, there are no operational tidal turbine farms of this type. But European Union officials have identified 106 sites in Europe as suitable locations for such farms. The Philippines, Indonesia, China, and Japan also have underwater turbine farm sites that might be developed in the future. The cost of such massive undersea structures is likely to be high. It is an open question whether future increases in petroleum costs will justify extensive exploitation of the tidal resource.

Ocean Thermal Energy is another potential source. The sun heats the surface waters of the tropical oceans, making them considerably warmer than water at great depths. It is possible to run a heat engine between these two thermal regions. A working fluid, such as ammonia, placed in a partial vacuum, is evaporated by heat taken from the warm surface water, the evaporated gas expands against a large turbine, making it spin to produce electricity. The gas is condensed after passing through the turbine by cooling it with deep ocean water. Since the temperature difference between the two heat reservoirs is modest, in comparison with a fossil-fuel steam power plant, the efficiency of conversion to electricity is quite low. On the other hand, the "fuel" (solar heated water) is free for the taking, so that such a plant should eventually pay for itself over time. So far, no commercial OTEC plant has been built, mainly for reasons of excessively long payback times. As energy prices increase, payback times shorten, generally leading to the opening of new markets.

Ocean Waves carry a large amount of energy. According to the US DOE, the total power of waves breaking on the world's coastlines is estimated at 2 to 3 billion kW. In favorable locations, wave energy density can average 65MW per mile of coastline.[15] Of course, due to environmental problems, land use conflicts, hazards to navigation, and for other reasons, only a small fraction of this power can be extracted for human use. Wave power devices extract energy directly from surface waves or from pressure fluctuations below the surface.

Wave power cannot be harnessed everywhere. Wave power-rich areas of the world include the western coasts of Scotland, northern Canada, southern Africa, Australia, and the northeastern and northwestern coasts of the US. Wave energy utilization devices have been built and operated in a number of locations around the world. Several European countries have programs to deploy wave energy devices.

Hydroelectric power generation is used extensively around the world. US hydro-power facilities can generate enough power to supply 28 million households with electricity, the equivalent of nearly 500 million barrels of oil per year. The total US hydro-power capacity – including pumped storage facilities – is about 95GW. There are probably a number of sites around the world where rivers can be dammed and hydro-power developed, but the environmental impacts can be huge. Thus, the potential for energy from this source is limited.

Geothermal energy comes from the interior heat of the earth. A fluid is heated below the surface, brought up, and used to generate electrical power. There are many ways this can be accomplished, and geothermal energy is already supplying power in many countries around the world. The inventory of accessible geothermal energy is sizable. According to the Geothermal Education Office,

> using current technology geothermal energy from already-identified reservoirs can contribute as much as 10% of the United States energy supply. And with more exploration, the inventory can become larger. The entire world resource base of geothermal energy has been calculated in government surveys to be larger than the resource bases of coal, oil, gas and uranium combined. The geothermal resource base becomes more available as methods and technologies for accessing it are improved through research and experience.[16]

It is possible to access geothermal energy anywhere on earth, but in most places it lies very deep. In special geological areas, however, it is closer to the surface and more economically available.

If we combined the energy production potentials for all the sources mentioned above, we could supply all the world's energy needs, probably several times over. There are limits, however, to this ambition.

PROBLEMS WITH RENEWABLES

Though promising, renewable energy sources are not unlimited, nor without their environmental impacts. At the current minuscule level of renewable energy generation, what little environmental consequences might result from renewable systems is pretty much a drop in the bucket compared with those of fossil fuels. However, as fossil fuels switch roles with renewables, the relatively minor impacts experienced now may grow to a substantial size. MTI describes some of the problems with renewables:

> Despite their benefits, renewables present important issues that need to be addressed. The main constraints on their use are the costs of the energy they produce and the local environmental impacts of renewable energy schemes. Currently, the cost of energy from renewables is generally higher than that produced by "conventional" energy sources. However, as renewables become more established and the benefits of mass production take effect, the gap will reduce. Indeed, in the case of wind power and some other technologies, this is already happening.[17]

As energy prices rise with the decline of oil, the cost-effectiveness of the renewable options should increase. The *impacts* of renewable energy technologies, I believe, can only increase. These impacts are many and varied.

Wind Turbines have been shown to be hazardous to birds in some locations. Though the problems are relatively minor at present, if the landscape is covered with these large devices, more problems can be expected. The few wind farms presently in operation are something of a curiosity, and generally well tolerated by most local residents. If the areas covered by them increase 100-fold, however, opposition on visual and amenity grounds can be expected to increase. Some of the turbines are noisy. This can be minimized with better technology, but noise pollution is likely to remain as an impact on local human settlements.

Offshore wind turbines have some advantages, mainly through being distant from the populations they serve. Out of sight means out of mind, unless serious problems develop. The hazards due to offshore wind farms include dangers for navigation and the disturbance of local marine fauna. When human structures are placed near the coastline, they are often heralded as "artificial reefs," capable of increasing populations of a variety of marine species – generally considered a good thing, but a possible problem when huge areas are involved.

Undersea Current Energy extractors have much potential in regions where conditions are right. They do, however, slow (and possibly redirect) the currents from which the energy is extracted, with possibly adverse consequences for marine life and for human shorelines, if the changes lead to significant alteration of geological features and habitats. For example, if a sufficient number of current energy extractors were placed across the Straits of Florida, the Gulf Stream's surface waters would be slowed by some amount, possibly altering the flow of this warm current, which is partly responsible for moderating the winter climate of Northern Europe. Though some undersea turbines have been designed to turn at relatively slow speeds – slow enough for marine animals easily to avoid collision – extensive experiments to validate this claim have not been completed.

The environmental consequences of slowing the Gulf Stream by a significant fraction, even if only at the surface, and the enormous cost which would be inherent in such an effort, would clearly not be justified by the relatively small part of Florida's future electrical energy use that could be supplied from this source. (Florida is growing at a rate of 2.35 per cent annually, corresponding to a doubling time of 30 years. That is a faster rate of growth than Haiti, at 1.73 per cent, India, at 1.8 per cent and Mexico, at 1.95 per cent.) Extraction of energy from other ocean currents may be more feasible, but the potential environmental impacts could be very large. The resource is so huge, however, that limited use of ocean currents will probably become desirable, as we search for alternatives to the fossil fuels.

Tidal Energy also has a large potential, but is restricted to estuarine areas experiencing significant tidal swings. Tidal power plants that dam estuaries can impede sea life migration, and silt build-ups behind such facilities can impact local ecosystems adversely. Tidal "fences" may also disturb sea life migration. Newly developed tidal turbines may prove ultimately to be the least environmentally damaging of the tidal power technologies, because they don't block migratory paths; however, the future economic feasibility of these huge underwater structures, anchored to the bottom, has not yet been proved.

Ocean Wave Energy has a very large potential, but also many environmental and technological hurdles to overcome. Impacts include the following:

- Hydrological effects of structures could alter the shoreline and adversely affect shallow areas, and the plant and animal life in these areas.

- There are potential navigation hazards. This might be mitigated with proper signaling devices, such as reflective paint, radar reflectors, and sound sources, but this hazard would remain.
- Some devices can be very noisy. The potential for damage to marine mammals is relatively unknown, but many species use sound waves for a variety of communication purposes. For humans, this problem is likely to be little more than an annoyance.
- When located on or close to shore, significant visual effects are likely.
- Some recreational uses of affected areas may be impacted, in some cases significantly.
- The installation of ocean wave energy conversion devices and the laying of electrical cables will damage and affect species on the sea bed and in the water column.
- Marine mammals will also be affected in several ways during the installation, and possibly in the operation, of devices.

The environmental consequences of extracting substantial fractions of the wave energy incident upon a shoreline can be very significant. The natural processes involved in beach erosion and replenishment are many and complex. The placement of a wave machine directly in or just offshore from the surf zone of a beach could have drastic consequences for beach formation dynamics, habitat destruction, and recreational use. Offshore sites may not have the same impacts, but if the wave regime approaching a coast is altered significantly there could still be serious effects. Wave energy system planners will need to choose sites that preserve scenic shorefronts and avoid areas where wave energy systems are likely to significantly alter sediment flow patterns on the ocean floor and littoral drift along the shoreline.

Hydroelectric Power has a number of potential environmental impacts. In addition to the flooding of valleys and destruction of upland habitats, hydro-power technology can have additional environmental effects, such as fish injury and mortality from passage through turbines, as well as detrimental effects on the quality of downstream water. A variety of mitigation techniques are now being used to address these environmental issues, and environmentally friendly turbines are under development. The flooding of large areas of land, however, constitutes a conversion of that land from its current use to an aquatic ecosystem. This has obvious societal impacts, and serious ecological ones as well.

Geothermal Energy seems a relatively clean option, but there are some environmental and social impacts. According to the Geothermal Education Office:

> Hydrogen sulfide gas (H_2S) sometimes occurs in geothermal reservoirs. H_2S has a distinctive rotten egg smell that can be detected by the most sensitive sensors (our noses) at very low concentrations (a few parts per billion). It is subject to regulatory controls for worker safety because it can be toxic at high concentrations. Equipment for scrubbing H_2S from geothermal steam removes 99% of this gas.
>
> Carbon dioxide occurs naturally in geothermal steam but geothermal plants release amounts less than 4% of that released by fossil fuel plants. And there are no emissions at all when closed-cycle (binary) technology is used. Geothermal water contains higher concentrations of dissolved minerals than water from cold groundwater aquifers. In geothermal wells, pipe or casing (usually several layers) is cemented into the ground to prevent the mixing of geothermal water with other groundwater.
>
> No power plant or drill rig is as lovely as a natural landscape, so smaller is better. A geothermal plant sits right on top of its fuel source: no additional land is needed such as for mining coal or for transporting oil or gas. When geothermal power plants and drill rigs are located in scenic areas, mitigation measures are implemented to reduce intrusion on the visual landscape. Some geothermal power plants use special air cooling technology which eliminates even the plumes of water vapor from cooling towers and reduces a plant profile to as little as 24 feet in height.[18]

This brief review shows that, despite the attractive energy potentials of the world's wind, waves, ocean currents, tidal currents, and geothermal sources, extracting significant portions of this energy presents environmental, economic, and social problems. Fortunately most are relatively easy to overcome when the power plants are small or widely separated. Modest use of these renewable technologies may be easily tolerated. However, considering the growth in world population and rising expectations for plentiful energy, future demand will put great pressure on developers to capture as much of each resource as possible, perhaps with terrible environmental consequences. A judicious use of all the profiled renewable energy sources should be able to meet the energy needs of a smaller world population with minimal environmental impact, thereby making the goal of a fully sustainable society realizable.

CONSERVATION LIMITATIONS

Energy conservation is another important strategy – almost like having a new source of energy, because energy freed up from one use is available for others. There are a number of technologies for conserving energy, including such things as compact fluorescent lightbulbs and energy-efficient buildings (and their efficient heating, refrigeration, and cooling equipment, as well as more energy-efficient appliances). The potential for energy savings through technologically-based conservation methods is deemed huge by thoughtful analysts. Improved energy efficiency in transportation systems is also possible, but not without massive redesign of transport vehicles and systems.

Under any hypothesis, efficiency must be a component of any strategy for meeting future energy needs. With the exceptions of ocean waves and ocean currents, solar energy in all its forms is a relatively dilute form of energy. The equipment needed to collect and convert it into a more useful form is generally expensive and burdened with serious environmental impacts, when deployed on a large scale.

Energy conservation is a very important adjunct of any solar energy conversion technology. Reducing one's energy needs also reduces the size of solar energy system needed, and hence its cost and environmental impacts, making solar a more socially and economically viable option. When you reduce energy consumption considerably through efficient design and construction, capital costs for a solar energy system needed to meet the reduced energy needs will be reduced as well.

The problem is that efficiency by itself is just a multiplier. As pointed out to me in a conversation with my colleague, Paul Jindra, no matter how "efficient" our energy-consuming processes are, and in complete compliance with the first and second laws of thermodynamics, a finite earth still gets consumed. With continued growth of population and affluence, and with overall consumption outstripping energy conservation, our energy and environmental crises are sure to remain with us for a long time.

CONCLUSIONS

It is clear that attempts to solarize the world economy are fated to run into serious obstacles, unless population and per capita energy consumption are drastically reduced. A major commitment to solar energy is likely to transform landscapes and seashores, bringing forth many new environmental problems, while demanding very large capital spending. My point is not that renewable energy technologies cannot work, or that they will be too expensive; it is that our thinking is flawed if we ignore the environmental and other adverse consequences of "solarizing" a growing industrialized world. Just because renewable energy resources are abundant, this does not mean we can grow our populations and economies indefinitely using only renewables, and still avoid serious environmental and economic consequences.

Renewable energy cannot be considered a complete panacea for all our energy problems as long as world population is not reduced and worldwide per capita demand keeps rising. There is a fallacy in believing that energy conservation and solar energy alone can save our energy future. Stuart Gleman once said that "solar provides just enough," but not enough that we can be wasteful with it.[19]

With a reduced population and lower per capita demand, energy conservation and solar energy should be sufficient to support a sustainable and sizable human population. Further, conservation and solar energy are needed now to conserve the valuable, complex molecules of fossil resources for more important uses later. Every country in the world should be funding research and demonstration projects, and promoting solar energy and conservation vigorously. Unfortunately, most are not pursuing this work vigorously enough. One reason is that many renewable energy technologies are perceived as not economically competitive with fossil fuels. In some cases this is true, but renewables have proved their worth. In many cases they are already a cheaper source of energy than certain fossil fuels.

Economic analysis and financial costing remains tilted against renewable energy; renewables are still held back. In the US, government subsidies to the fossil-fuel industry are not matched for renewables. Existing electricity systems and laws often make it difficult for renewables to gain access to national markets. Fossil and nuclear power in the EU are publicly subsidized to the tune of 15 billion euros per year. In addition, the European taxpayer picks up the environmental and human health bill for acid rain, for NO_X emissions, for particulates, and for the "natural" disasters caused by climate change, itself triggered by fossil-fuel burning.

Perhaps the most serious problem is that current "free market" practices include no mechanism for including non-monetary benefits along with the energy savings of a given technology. Such technologies are therefore undervalued in the market, and under-utilized as a consequence.

Despite these obstacles, the market for renewables is growing. The jobs, income, and energy security that will result from building market dominance for renewables are undisputed. The only question is how fast it will happen. If world population and per capita use of energy continue growing at current rates or higher, our demand for energy is likely to grow faster than our ability to supply it from renewable sources. Also, as supplies of non-renewable sources dwindle it will become increasingly expensive to supplement solar energy or back it up with non-renewables. This reminds us of Bartlett's Second Law of Sustainability: "In a society with a growing population and/or growing rates of consumption of resources, the larger the population, and/or the larger the rates of consumption of resources, the more difficult it will be to transform the society to the condition of sustainability."[20]

It seems clear that the industrial nations of the world, as well as those working to industrialize (mainly along the Western model), must implement policies to stop population growth, reduce per capita energy demand, conserve valuable fossil resources for more important uses, and aggressively promote the use of renewable energy while working hard to reduce environmental impacts. There are indeed limits to growth, and the human species is now approaching them. In order to achieve the goal of a sustainable global human society, considerable education, discussion, and rethinking of priorities will be necessary. Let us begin.

12
Population, Energy and Economic Growth: The Moral Dilemma

Ross McCluney

Before the fossil-fuel-driven Industrial Revolution and the improved transport of foodstuffs, goods and persons that this enabled, population density correlated with resource availability. A too-large population could not long live beyond the limits of the local ecosystem's carrying capacity. As transportation systems advanced due to the ever-increasing availability of cheap fossil-fuel-based energy, physical and biological carrying capacity limits could be exceeded in a region by importing resources from other regions. This can continue for as long as energy remains cheap and abundant, and while environmental and economic impacts remain tolerable for human populations.

The process has now been carried to extremes; our world's human population exceeds the physical and biological carrying capacity of the whole earth, made possible solely by fossil fuels. Wackernagel and Rees developed the concept of an *ecological footprint*, the biologically productive land or sea area required to produce sufficient resource yields for the supported human population, and to absorb the corresponding carbon dioxide emissions.[1] The same method is presented in the US National Academy of Sciences publication, *Tracking the Ecological Overshoot of the Human Economy.*[2]

Redefining Progress produced a November 2002 report outlining the ecological footprint of 146 nations. As Mathis Wackernagel, the Sustainability Research Program Director, notes: "Humanity's ecological footprint exceeds the Earth's biological capacity by about 20 percent," continuing, "many nations, including the United States, are running even larger ecological deficits. As a consequence of this overuse, the human economy is liquidating the Earth's natural capital."[3]

About three quarters of the world's current consumption of resources is by the approximately 1.2 billion people living in what are called "rich nations," while the remaining quarter is consumed by the other 5 billion people currently living on this planet – almost a third of whom are categorized (by various UN agencies and other organizations) as living in great or extreme poverty. It would require at least three times the earth's entire resources and physical area to provide all the world's current population with the material and energy currently consumed by an average North American citizen. This immediately leads to questions not only on the physical possibility, but also the logical validity of economic expansion as a remedy for poverty. Since oil and natural gas drive all types of classic or conventional economic expansion, the peaking and subsequent decline of world oil production poses immediate and direct challenges to attempts at achieving unlimited global economic growth.

Geological consultant Walter Youngquist writes,

> That oil production will peak and then decline is not debatable. If the more optimistic are right, and the peak date is a little further away than most geologists now predict, this would simply exacerbate our problems, for it means that the population at the turning point of oil production will be even larger than it would be at an earlier date, and it will then be more difficult to make the adjustment toward life without oil. Envisioning what the post-petroleum paradigm will be like involves consideration of myriad facets of the world scene. The worldwide decline of oil production, ultimately to the point where it is insignificant relative to demand, will have many ramifications, changing world economies, social structures, and individual lifestyles.[4]

The steady decline of world oil production after Peak Oil and over the next few decades (with a 25 to 30 per cent reduction from today's production by 2025 being probable) makes for a somber scenario regarding world food production and food availability per head of population: a significant decline in world population, due to this single factor, is more than possible. How we adjust as a global community to this challenge can only be of great concern to us all.

Proponents of staying on course – that is, "business as usual" – say that as we run out of oil it will be replaced by extreme energy conservation and a radical switch to renewable energy sources. At the same time, political leaderships resist any significant increase in oil and energy prices, while at every moment declaring their faith in "market mechanisms." Though technology improvements – but *not* breakthroughs – are possible and likely, few signs exist today that intensive energy conservation and a switch to new and renewable sources of energy is taking place, or is even being coherently promoted by world leaders.

The growing availability of cheap oil over the last century and a half has led to an enormous expansion of human population, industrial and technological impacts on the environment, and extreme dependence in the urban industrialized nations on cheap fossil energy. So long as cheap energy subsidizes and enables bulk transport of vital raw materials, food commodities, energy minerals and industrial goods, and as fossil fuels decline and energy becomes inexorably more costly, humanity will reach a turning point. The current course cannot be continued indefinitely, and by this I mean *for more than a decade*.

RESOURCES AND POPULATION

Resources and goods are traded globally. A result is that regions with inadequate supplies of any input can make up for it with resources imported from elsewhere, as long as their demand is solvent (that is, if they have the cash for importing resources). If the raw materials for a factory are not locally available, they can be and are shipped around the world to where they are needed. If the oil products, natural gas or electricity needed to run large transportation systems are not locally available, large pipelines, oil and gas tankers and electric power grids will transport the energy to where it is used, and over great distances. If the soil is too poor, and fresh water inadequate to grow crops well in a region, fossil-fuel-derived or powered machines, fertilizers, pesticides, soil conditioners, irrigation pumps, and so on, will be used to maintain output, or even temporarily increase agricultural productivity. Increased food production usually supports larger populations, and these populations depend upon imported, external resource inputs, fuelled by cheap oil and other fossil fuels to maintain themselves.

The most important resource, and the single biggest item of world trade in volume terms in the so-called global market society, is petroleum. Table III.12.1 gives imports and exports of crude petroleum for selected countries, along with their populations, ordered by population size. The largest importer in the world is the US; second is the relatively small Japan, having just under half the imports of the US. The combined population of the US and Japan is 414 million. The largest exporter is Saudi Arabia, with a population of only 24 million, but increasing at 3.8 per cent each year. Saudi Arabia, the United Arab Emirates, Iran, and Iraq export a combined total of oil equal to the imports of the US and Japan, but with the exporters having only 16 per cent of the importers' population.

Table III.12.1 World petroleum supply and disposition, 1999
(crude oil only)

Country	Primary Supply Crude Oil Imports (Thousand Barrels/day)	Use of Supply Crude Oil Exports (Thousand Barrels/day)	Net Exports over Imports (shown by "+") (Thousand Barrels/day)	Population in Millions
China	745	144	−601	1,281
India	874	47	−827	1,049
United States	8,731	118	−8,613	287
Brazil	483	1	−482	174
Russia	91	2,648	+2,557	143
Nigeria	0	1,834	+1,834	130
Japan	4,223	0	−4,223	127
Mexico	0	1,580	+1,580	102
Germany	2,118	35	−2,083	82
Philippines	325	0	−325	80
Egypt	0	284	+284	71
Iran	0	2,531	+2,531	66
Thailand	698	16	−682	63
United Kingdom	744	1,729	+985	58
France	1,673	0	−1,673	60
Italy	1,628	0	−1,628	58
Korea, South	2,406	0	−2,406	48
Colombia	4	516	+512	44
Spain	1,163	0	−1,163	41
Poland	318	0	−318	37
Algeria	5	744	+739	31
Canada	836	1,059	+223	31
Venezuela	0	1,923	+1,923	25
Iraq	0	2,025	+2,025	24
Saudi Arabia	0	6,514	+6,514	24
Korea, North	44	0	−44	23
Taiwan	740	0	−740	22
Netherlands	1,094	6	−1,088	16
Chile	184	0	−184	16
Ecuador	0	236	+236	13
Angola	0	699	+699	13
Greece	319	2	−317	11
Hungary	114	0	−114	10
Sweden	405	0	−405	8.9
Israel	218	0	−218	6.6
Libya	0	1,069	+1,069	5.4
Finland	219	0	−219	5.2
Singapore	889	0	−889	4.2
United Arab Emirates	0	2,009	+2,009	3.5
Kuwait	0	948	+948	2.3

U.S. Department of Energy
Energy Information Administration
http://www.eia.doe.gov/pub/international/iea2000/table31.xls

Population
Reference
Bureau

Though most developed countries in Europe and Japan have either fairly stable or declining populations, the US is still growing quite rapidly, mainly due to high levels of legal immigration, tolerance of illegal immigration, and the relatively high initial fertility levels of the new immigrants. US population at the current rate of expansion might attain about 350 million by 2050. In all cases, developed countries depend on drawing resources from less developed ones. The financial, political, and cultural aspects of this are described as "underdevelopment," which we might more accurately call *exploitation*. It is therefore not surprising that less developed countries retain a certain level of suspicion regarding anything the developed world might offer in the way of "aid and assistance." The problem is particularly acute for any concerted effort to control world population well in advance of fossil fuel depletion taking the decision out of human hands – through mass starvation. One aspect of the developed/ underdeveloped nexus is that less developed countries often resent "solutions" offered, or imposed, by the richer countries.

Fortunately, the United Nations – with most countries of the world as members, including both haves and have-nots – has a fairly aggressive strategy to reduce world population growth rates. Information, education, and materials are offered to help people limit family sizes and live well with the children they do have. Declining infant mortality alone goes a long way towards lowering fertility. With fewer infant deaths, parents do not have to have so many babies in order for one or two to reach adulthood. This is one way the developed world can provide assistance without being accused of telling the less developed countries what to do – by funneling aid through the United Nations, representing all countries. (This will not work, of course, if the developed world dominates UN population policies, or withholds funds to support them.)

The UN Population Fund (UNFPA) and the UN Population Information Network have active programs to help countries control runaway population growth and to assess its causes. When releasing the UNFPA report titled *The State of World Population 2000*, its Executive Director, Dr. Nafis Sadik, said: "Millions of women are denied reproductive choices and access to health care, contributing each year to 60 million unwanted or mistimed pregnancies and some 500,000 preventable pregnancy-related deaths. Nearly half of all deliveries in developing countries take place without a skilled birth attendant present." The World Bank estimates for 1998 that 1.2 billion people worldwide had incomes of less than US$1 a day – that is, about 25 per cent of the population in less developed countries. Some 2.8 billion existed in 1998 on less than US$2 a day – that is, well over one-third of the world's total population. At least half of the current population of the world lives in poverty.

ENERGY AND POVERTY

Both natural resources and wealth are unequally distributed. Even the so-called "rich oil exporter nations" are mainly low- or medium-income countries, using World Bank criteria (from $400 to $1,500 GNP per capita). Increasingly, as world economic rates decline, the growing dichotomy between the haves and have-nots leads to political instability, ethnic and community conflict, civil disturbance, and often armed conflict.

According to DOE projections,[5] growth in world energy use between 1999 and 2020 will reach a total of 52.4 per cent, or an average of 2.5 per cent per year, corresponding to a doubling time in world energy consumption of a mere 28 years. Population growth, industrial development and increasing per capita energy use are the major determinants of energy demand growth. This returns us to the crises that will surely come as we pass through Peak Oil and enter a period of ever-declining supply. No immediate solution is likely, given the current global economic and political context.

One of the central "riddles" of development is how commercial energy supplies, which enable and maintain conventional economic growth, can be "taken out of the equation" without chaotic economic impacts. All models of development are growth-based and growth-seeking – yet if all countries consumed energy at US or even European rates, the depletion and then final exhaustion of remaining oil and gas reserves would take only 20 or 30 years, instead of about 60 years. The environmental impacts due to a world population of around 9 billion persons consuming fossil energy resources at US or even European per capita rates of today would most certainly be catastrophic. The challenge therefore is to *reduce consumption* in the rich nations, and for poor countries and people to escape poverty without crippling their economies, the biosphere, climate, or other natural support systems.

The greatest challenge facing us is to improve the lot of the poor *without* greatly increasing the inefficient and polluting use of fossil fuels. Reduced consumption by the rich countries and energy conservation are two immediate options. Development of new and renewable energy sources is another. But these must be accompanied by the halting, and then reversal, of world population growth.

HOW MANY PEOPLE?

Joel Cohen wrote a well researched and scholarly book titled *How Many People Can the Earth Support?* He tracks the historical answers to his primary question from the earliest one listed, in 1679, to the year 1994. Estimates range from the very small (500 million people, by Ehrlich) to the ultimate extreme of a world inhabited by human beings at a density of 500 persons per square *meter* of land surface, in buildings with an outer skin temperature of 2,000°C, this "heat limit" world population being about 1 thousand-million-billion. Most variations in estimates made over time are due to varying levels of scientific knowledge (lower for earlier forecasters), and to widely varying assumptions regarding what physical or biological factors set final limits on human population.

In a recent book[6] I responded to the question of how many people the earth should support. My answer is that this will depend on the kind of world you want. Will all people live at similar standards of living, or will this vary widely, as is currently the case? What is the most important failure factor in human population growth? Is it loss of food, energy, or capacities for waste removal? Or does the limit come from the spreading of killer diseases? Since at least 30 years ago, we have added the possibility of global thermonuclear war annihilating all major (and even minor) centers of population.

David Pimentel of Cornell University wrote a short article estimating the maximum carrying capacity of the planet.[7] He made an assumption that all people would attain 1999 average standards of consumption in the US. His estimate called for a substantial *reduction* in the current human population, to about 1 or 2 billion, arguing that the earth would be incapable of supporting the current world population at American levels of affluence. This conclusion is similar to Wackernagel's – in other words, that supporting the world's current human population at US levels of affluence would require not one but *three* earths.

Virginia Abernethy commented on these estimates, as follows:

It is small wonder that numerous students of carrying capacity, working independently, conclude that the sustainable world population, one that uses much less energy per capita than is common in today's industrialized countries, is in the neighborhood of 2 to 3 billion persons. [We should also] note the congruence with Watt's projection of rapidly declining population size near the end of the Oil Interval. The absence of cheap, versatile, and easily used sources of energy, and other resources, seems likely to change the quality of human life and may even change, for many, the odds of survival.[8]

We are currently supporting a much larger population than Pimentel's estimate, because most of the current 6 billion people have a substantially lower standard of living than the one he used for making his projection. In addition, our *temporary* sources of energy (depleting fossil fuels) are supporting populations which, without energy-extended resource domains, would not be able to continue existing. Any estimate of long-term maximum supportable human numbers must assume the absence of substantial quantities of fossil-fuel resources (see Chapter 11). A substantial increase in human population above the current 6-billion-plus mark might be possible, but only: (1) at the expense of other life-forms with which humans compete, but on which humans also depend; (2) by lowering the overall material standard of living (and ecological impact) of the current human population; or (3) by finding ways to reduce human impact on other life-forms while human population and affluence continue to grow. The last of these alternatives can only be considered wishful thinking, but is grist to the mill of the "limit denial industry," the technological optimists, and growth-seeking economists believing in market-triggered "human inventiveness."

All estimates of large increases in the world's population are constrained (while generally denying this) in ignoring human rights, rejecting biodiversity, depreciating the aesthetic and cultural aspects of natural environments and non-human life-forms, and reducing human beings to mindless goods-consuming units. If we maintain the industrialized world while allowing the rest of the world to grow substantially in numbers, the consequence is to doom much of that "other world" to perpetual misery, while clinging on to the industrialized way of life as desperately and as long as possible. This scenario would very probably lead to a major population crash, as energy and other limits were quickly reached and then exceeded.

It is clear that we are facing serious moral and ethical issues as we approach the end of the "petroleum interval." To focus on the moral questions, I postulate three different scenarios:

Business as Usual Scenario. The industrialized nations continue as they are, increasing energy efficiency and switching gradually to renewable energy sources, even as their populations grow slightly, primarily from immigration, and their per capita energy consumption levels remain high. The underdeveloped nations continue with current trends of intensive agricultural development, urbanization and industrialization, using more energy and resources, and generating more pollution in the process. This continues until the combined stress on resource availability and the environment yields a catastrophic collapse of social and political systems. Worldwide economic collapse will most certainly be an encroaching condition in the process. The present conflict over resources, especially oil, escalates to the level of world wars. Through these wars and their impact on food production, and assuming they are non-nuclear (see Chapter 19), world population may be reduced drastically. However, "collateral damage" to the environment and resource-supplying systems will be large, precipitating a die-off. Several billion people will die in a short period, before sustainable levels are reached.

Selfish Nation Scenario. The industrialized nations, seeking to maintain their affluent ways of living and materialistic perspectives, isolate themselves as much as possible from the rest of the world. Sources of oil in weak, remote nations are appropriated by economic and (increasingly) military force. The less developed world is shut out and cut off from the benefits of affluent living. Powerful industrial nations try by all (necessarily military) means to perpetuate this extreme dichotomy. Populations of the less developed nations decline sharply, with the breakdown of political, social and cultural structures.

Humanitarian Scenario. A massive program to reduce population growth is initiated, and strongly supported, with financial and economic resources from the developed countries. Developed nations embark on a global program of radical resource efficiency and increasing utilization of new and renewable energy sources (NRES). All perfected NRES are made available, with financial and economic support, to less developed nations. Energy-wasteful transport and urban–industrial practices are rapidly and radically modified, to reduce energy consumption and environmental impact. World population stabilizes in a few decades, thereafter declining gradually to approximately 4 billion by 2050.

Which of these scenarios we will produce is a matter of vast importance. If we choose the last of them, considerable public education, debate, and enlightened decision-making will be required. In choosing amongst the possible future scenarios, we are facing serious moral and ethical decisions. The issues must be debated extensively, globally, and rapidly. Since humanity has taken over control of Spaceship Earth, we simply must decide where we want the great ship headed.

THE MORAL DILEMMA

Do the industrial countries owe anything to those in underdeveloped countries living lives of misery? Will the industrial world be willing to alter its own system to benefit the starving billions elsewhere? How much should the industrialized countries be willing to sacrifice for the sake of the underdeveloped world? Is it moral to conclude that we should not make such sacrifices, or is the very question born of a fallacious understanding of what it takes to live well? These are serious questions demanding thoughtful answers.

It is difficult to motivate people to change to a lifestyle they see as less desirable than the one they currently enjoy. I believe that ways can be found to live better with much less energy, and with lower material consumption, than currently enjoyed in the developed world. We can find non-polluting, low-energy ways to live, eat, sleep, dream and enjoy life. By slowing our pace and simplifying our lives, we can learn to live well and happily without destroying the earth which is our home. Positive visions for humanity's future are needed to provide the essential motivation for change. Under any scenario, population growth must stop – ideally by human instigation, to prevent nature from having to do it for us. It is in fact an open question whether even the UN population projections are feasible and possible *without* bringing the world to the brink of near extinction, because of excessive damage to many critical components of the natural life-support systems of the biosphere. Finding out if this is possible is an experiment we are now pursuing. My children, and those of any parent, will see how it comes out.

13
Apocalypse 2035

Andrew McKillop

By no later than 2035, but in a vastly different world from today, many of the energy and environment limits discussed in this book – most of which are still the subject of a denial industry today – will have been proved. By 2035, oil and gas production, and therefore consumption, will have fallen by as much as 75 per cent and 60 per cent, respectively, from today's levels. Coal production and consumption may well have bounded upwards – but if so the environmental and climatic consequences will be grave. Where coal can be used as a substitute for oil and gas (often with large process or conversion losses of energy), the depletion period for this coal burn will dramatically shorten the lifetime of coal reserves from the levels of "up to 400 years" that are touted today, to perhaps no more than 50 years (see Part V). Radical energy solutions such as the entirely theoretical hydrogen economy (based on huge programs of nuclear reactor construction to produce hydrogen by electrolysis, coupled with intensive energy conservation, and very large-scale use of solar and renewable energy sources) is predicated upon massive investment, or root-and-branch economic restructuring and extensive lifestyle changes. None of this is likely to come about voluntarily. Because of economic, social and political inertia, it is more likely that fossil-energy sources will dwindle rapidly, leaving only stark choices. World climate, by 2035, will have substantially changed from today; the effects of carbon dioxide levels not seen for 400,000 years will most certainly wreak major, accelerating, but at present unpredictable changes in climate and sea levels. Species extinction due to the human footprint will by 2035 test all of mankind's ingenuity with genetic manipulation of those plant and animal species decreed useful and necessary for building and operating an urban–industrial "heaven" on earth.

The period of unfettered economic growth of 1950–80 will probably seem a mythological belle epoque to surviving consumers in 2035: political leaders of that time may claim that it is possible to recreate it, but inherited or leftover energy systems from that period will themselves present enormous risks. By 2035 hundreds of nuclear reactors of the first and heroic age of nuclear weapons and energy – which gave the world Hiroshima, Chernobyl and not-so-cheap electricity – will have been decommissioned. Some will receive skilled and costly attention in separating, processing and storing deadly nuclear materials far from any contact with the biosphere. Many others will have been decommissioned as Russia has done with its submarine fleet's nuclear power reactors since its entry to the global market: dump them in the sea. Others will probably be operated too long, to squeeze just that bit more "cheap" electricity from them, suffering grave structural faults due to neutronic aging, and explode like Chernobyl. Human societies will have had to make the decision to go on – or not – with the nuclear experiment. Even in 2003 the number of nations which have renounced nuclear power is growing, but after Peak Oil and the end of cheap fossil energy it is uncertain that such courageous decisions will stick, unless and until spectacular nuclear reactor accidents remind consumer-citizens of what they depend on for cheap electricity. This is simply because, regardless of their refusal and denial of limits, human beings also love their creature comforts. An ace card for the limit-denial industry is none other than nuclear power itself – a basis for propagating images of an electric-and-hydrogen New Jerusalem for every faithful, determined and unyielding member of the consumer community. With every dollar price rise for a barrel of oil, the nuclear lobby gains voices, but the *wipeout* of cheap and abundant reserves of oil and gas will be so fast that fantasy projects for building New Jerusalem's vast infrastructures will almost certainly be overtaken by events. While promoters of the nuclear option say that inaction and hesitation can bring about an apocalypse, inaction and inertia in the present will eliminate options for maintaining consumer civilization. Consumers of this present generation may be the last who can casually live "high on the barrel," and pretend they have no reason at all to contemplate real and serious change.

UNSATISFIED EXTERNALS

The word "apocalypse" is from the Greek for "revealing." Apocalypse has its somber meaning because of St. John's Apocalypse. St. John was one of the twelve apostles, a contemporary of Jesus Christ, and said to have witnessed the empty tomb of Jesus, following his burial after crucifixion, thus "proving" that Jesus had risen to heaven. St. John is said to have written his Apocalypse in about AD 94 on the island of Patmos, where he was exiled by the Roman military province of what is today Israel and Palestine. Legend has it that St. John, condemned to death for his fiery predication by the Romans, had been boiled in oil, but miraculously survived. At the time, the outlook for Christians was very bleak. His "chosen people" had been dispersed, persecuted, their prayer meetings banned, and Jesus crucified in public. The Roman Imperial war machine rolled on. Christian traders had in some cases been marked or branded (on their hands, arms or foreheads) and forbidden to trade, while other cults and sects in the region were tolerated, and had large and sometimes wealthy followings. St. John's Apocalypse, therefore, was both a rallying cry and a revenge on existing political power and competing religious cults.

It thus reflects the context of oppression and despair for the leadership and members of a new and emerging cult in open competition with established myths and beliefs, and describes in lurid detail what will befall those who oppress and deny Christians. Only by total destruction and renunciation of the works, ways and beliefs of those other communities worshipping what Christians considered inferior, salacious or bestial religious icons, said St. John, would survivors who adopted the true gospel of Jesus be able to live in wealth and peace in the New Jerusalem – Heaven on Earth – that he described in poetic detail. His Revelation or Apocalypse was heroic in its technological details; the New Jerusalem he described was built on foundations 144-elbowlengths (about 60cm) thick, of jasper, sapphires and emeralds, needed no street lighting because nighttime was banished by dispensation of God, and the city was 144 million square stades in area (one stade equals 60 Greek feet, or about 157 meters). It was lined with trees constantly bearing fruits and nuts, had crystal clear rivers running through it teeming with fish; home to 144 million faithful and contented citizens. Like any pamphleteer seeking to persuade readers that there is not a moment to lose in a race between Good and Evil, between Paradise or Hell on Earth, St. John used more than a little creative license. He targeted Hebrew and Greek myths, doctrines and beliefs – the Hebraic Abaddon and Greek Apollo as the most intensely evil, destructive cult figures. In particular, his literary invective was directed at Apollo, who he called "The Terminator" or "Destroyer."

When St. John was writing, the cult of Apollo was at least 2,250 years old. Further, the Greek cult of Apollo had changed with the social, economic, demographic and environmental developments in Greece and the Eastern Mediterranean – and had therefore also changed in its cultural meaning and role (see Chapter 23). In early Greek myth, before about 1500 BC, Apollo had indeed been very cruel, capricious, pleasure-seeking and orgiastic, and was the first Greek god to be openly homosexual. Conversely, he was never exclusively destructive, and Apollo cults were from early times associated with waxing and waning, sprouting and withering, spring and autumn, giving rise to multiple related mythical entities driving the seasonal changes in the biosphere – such as Thallo and Carpo, Auxon and Hegemone.[1] The celestial downfall of Apollo, however, was a farcical, even derisory affair of him going too far and attaching the wife of Zeus, Hera, to the sky with anvils tied to her ankles. To make matters worse, his son Asclepius – the first doctor – went too far in his experimental healing by resurrecting a dead person, and cheating Hades (the god of Hell) of a sure recruit. This resulted in punishment by the supreme god, Zeus, with banishment of Apollo for one year as a shepherd, a role that Apollo came to love. After a year's sabbatical Apollo was a very different god, preaching moderation in all things, and self-knowledge as the fundamental key for both gods and mortals, besides the good practices in sheep husbandry that he had devised. Interestingly, "The Lamb" was a synonym used by St. John for Jesus in his role of returning to preside over New Jerusalem, in the wake of nearly unlimited destruction and death following from the people's foolish, blind and greedy following of the ways of The Terminator.

As in Greek myths, uncontrolled – and at that time unexplained – environmental change was extensively used by St. John to luridly evoke the punishment his God reserves for wrongdoers. Using complex numerological associations (perhaps incorporating coded messages to readers outside his Patmos Island prison), St. John described floods, desertification and insect plagues of giant Hollywood-style locusts, as certain disasters for non-believers and those rejecting Christian ways and beliefs. In this he was drawing on public concern and fear: by the first century AD there had indeed been many "unexplained" environmental changes in the Eastern Mediterranean due to increasing population, urbanization, single-crop agriculture, deforestation, intensive grazing, irrigation and waterworks, and the modification of shorelines for ports and maritime facilities. Major floods had occurred in the region for more than a millennium, and had been incorporated into many myths, cults and religions. St. John's Apocalypse sensationally intensifies and distorts such side-effects of human impacts on the environment, biosphere and climate, while offering a shiny, gold- and jewel-bedecked, immense city, with a permanent agricultural surplus within its walls, as the reward for true believers.

DIVINE LEADERS AND CHEAP OIL CRUSADERS

Christianity, as we know, is the nearest thing to an "official religion" of the supposedly secular consumer society that current leaderships of advanced industrial societies defend with all the verbal, economic, and military power they can buy with taxpayers' money. For perhaps 20 years, Sunday television has been rife with footage of these leaders entering a church. Whenever they make war, which they do only "to defend civilization," according to the speeches their scriptwriters concoct, they predict suffering, destruction and death for non-believers in, and opponents to their own fossil-fuel-based New Jerusalem. Their version is really no different from St. John's imaginary "eternal" city – an immense expanse of stone and paving, vast thoroughfares, and buildings that reach into the sky, with any animal or plant other than the select few decreed as "useful" being banished to the barren places beyond. Non-believers in shopping mall *culture* are regarded as foolish, backward or primitive, while opponents are given exactly the treatment that St. John's Roman military oppressors applied to him.

Under any hypothesis for fossil fuel and environmental limits, current consumer civilization has much less than 50 years before its terminal crisis. As with the collapse of any civilization this will – surely – be an "apocalyptic" event; but those punished for predicting this end will almost certainly have to be boiled in vegetable oil, not petroleum, for their sins.

Part IV

Partying on in the Growth Economy

The title of this section is taken from Richard Heinberg's excellent book, *The Party's Over* and, apart from Colin Campbell's droll and cutting open letter to the US Geological Survey (USGS), all the chapters in this part are by myself. The focus of Part IV is the dominating pervasive myth in our society: that economic growth is always good, always possible, and if absent will always return. In 2003–04, every artifice and possible stimulant will be applied to the US economy to secure the re-election of George Walker Bush, of course accompanied by ever-more glaring "positive spin" on the official economic statistics his administration puts out. From that perspective, oil shock is about the most awful, catastrophic thing that could happen. Nevertheless, the apparent vintage growth of the US economy in late 2003, which was in sharp contrast to officially admitted recession in Japan and most Eurozone countries (including Belgium, Italy, France, Germany and Holland), also required a 22 per cent rise in the trade deficit on oil. In other words, US oil imports were up 22 per cent in terms of their dollar costs in 2003 over 2002, reflecting an apparent spurt of economic growth at an annual rate of 7.2 per cent in the third quarter. Inflation tracked this closely, also at over 7 per cent on an annual basis, and the US financial press lost no time proclaiming the culprit: high prices for oil and gas. An oil shock is therefore the dark nightmare of any Bush team adviser or analyst, as for politicians anywhere else in the rich world democracies, which is yet again curious. Economic history and reality show clearly that previous oil shocks have always triggered *bigger and better upswings* on major bourses, within at most 18 months from the end of each crisis. However, 18 months after an end-of-year oil shock in 2003 would have been an awfully long time to wait for the Bush team.

The founding era of the near-millennium fantasy that oil shock means price rises (that is, inflation), which means economic crisis, was in 1979–81. In the early 1980s the pervasive, self-reinforcing desire for cheap oil as our civilization's birthright, and a one-way ticket to economic prosperity through strong, non-inflationary economic growth, became firmly embedded. Since then it has had powerful downstream impacts on our collective psyche. Running out of cheap energy, after the cultural revolution inaugurated by the Thatcher–Reagan era at the start of the 1980s, would have been too much of a shock. As a direct result, inflation and oil shock were banished forever, together with the merest idea of physical scarcity for any cheap commodity, from the official mindset. Today, oil shock might grudgingly be admitted as a possible worst-case scenario, due to desert tyrants and despots able to ignore opinion polls, but not yet as the inevitable outcome of shrinking supply and unbelievable growth of oil demand for the no-longer-fantastic car fleets of China, India, Iran and other big, new industrial powers. All kinds of new economy myths are on hand to help our democratic leaders deny reality. The stock of myths they delve into has accumulated from the early 1980s, and is a little long in the tooth, though that has little or no importance since, after all, the New Economy is founded on the thoughts of such luminaries as Adam Smith, writing in the 1760s. Above all – and no political leader will miss out on this in their speeches – the party must go on.

Growth must be maintained, improved and consolidated whenever it isn't simply being increased. In the real world growth means more consumption of everything, and anything physical or material needs energy. This stepwise logic may be too complex for a standard-issue democratic leader, or even his spin doctor, but previous themes in this book have already cast more than reasonable doubt on the physical, geological and biological resource bases of the growth economy holding up for even ten more years. The reality gap therefore starts right here, and is exactly the same as that facing the New Economy trailblazers of the early 1980s: How do we get the cheap oil and gas to keep the party going? Early 1980s leaders waxed lyrical on imminent quick fixes like "clean and cheap" nuclear power, synthetic oil, or the dematerialized, downsized and delinked energy-lean economy, which singly or in combination would surely rout any price rises for oil. Today's clones and clowns, when not sending troops to Central Asia, the Middle East or West Africa to safeguard oil supplies and future democracy in a single operation, have only such unconvincing things as windmills, "demand destruction," and the Hydrogen Economy, to brandish like garlic at a vampire. Apart from the complex, even chaotic and contradictory ideas underlying the Kyoto Treaty, vaguely promising reduction in oil and gas burning, there is no way out today. No solution exists for the demise of cheap oil, or cheap gas, and so they cannot decline. The end was spotted, and so it was declared an optical illusion. End of discussion. Such hermetic reasoning, we can note, is characteristic of schizoid and paranoid conditions in individuals, but when they become mass society's "answer" to oil and gas depletion, we can be sure that it is real, and is on its way.

Several of the investment community's guiding lights only burn dim yellow, like bulbs in a US, Italian, British, Chilean, Japanese or Argentine electric power system brownout, when asked to comprehend the difference between technical shortfall and Peak Oil, which signals long-term and final resource depletion. The question is avoided, if necessary by incomprehension of the most educated sort. Furious work on paper archives, to discover growing reserves of paper oil through imaginative forecasting techniques – latterly much used by the USGS, as Colin Campbell explains – now feeds this optimistically contrived news to the analyst community tracking the fundamentals of our future growth and prosperity. When BP announced in early 2001 that it was changing its stripes and shedding its skin to become *Beyond Petroleum*, it first denied this had anything to do with depletion, and not long after denied it was changing its name, keeping that new identity only for solar electric cell publicity! Real company-owned reserve and production bases for major players in world oil, and even the numbers of its Seven Sisters sorority,[1] have inexorably shrunk. There are now five anxious dwarfs instead of Seven Sisters, and their own-company oil production capacity shrinks every year.

Like its political masters, the financial community exists to reject and "refute" reality, to keep the party going. Consequently, if supply shortage threatens, stock exchange gurus will revert to type and descend into the swirling cauldron of bourse cosmology as if for rebirth. Here they learn one thing they can never forget: shortfalls by definition are temporary, because new supply will always enter the market. A shortage of anything is just an *opportunity* for entrepreneurs – even if the only real opportunity remaining is practicing the world's oldest profession. In short, the challenge of scarcity and shortage calls forth all that is best in every person. This is proved by what these pop-eyed gurus call History, with a very big capital H. Even worse, the baleful non-science of economics exists for building a thick veil of flimsy logic around such semantic shell games and glaring tautologies. Through a free flow of news-management techniques operated between the business community and political leaderships in the global market world, there will always be political clout behind New Economics, assuring one and all that any shortage is, by definition, temporary. Dips in the index can only be temporary, and themselves are an opportunity to buy in cheap and be well-placed for the certain rebound. When or if stock exchanges are *temporarily* morose (as most were through 2000–03), there are a bunch of strategies to shift back and forth between interest-rate-related investments and stocks, while the equity markets are down. Only in extreme situations (as since 2002) should barbarous "relics" like gold be purchased, unless of course your favorite guru already had an inspiration that a new bull run was going for gold.

This laughable logic also applies to oil and energy. Thus if oil prices rise enough, in the shadow of that prowling bear some gurus will root for a host of Miracle Energy Incorporateds offering magnetic levitation and perpetual energy machines. Maybe 99 per cent will never make a dime, but hope can be creative, and hope is comforting to believers in our civilization because it shows readiness to innovate, and is proof that we can all depend on individual greed and egotism, or at least on chance and luck – the defining variables of our consumer civilization. Even in the 1980s, after the shaking world bourses took from Reaganomic adjustment to the Second Oil Shock, and more surely than ever in the 1990s Clinton Boom, bourse players and gurus, commentators and contented investors kept their faith in those powerful *adaptive responses* that the market stands for in the hearts and minds, or at least the imagination, of the faithful. Energy and oil shortage, therefore, can only be transient, at least in officially approved thought. Supply will rebound, perhaps at a higher price, but it will never shrink or fail. Unfortunately, Mother Nature played along with this fantasy in the early 1980s when this belief was given bulletproof protection around its mindless core. For a short while new oil and gas supplies *were* abundant, and they were located in the heart of Europe, not so far away from several major European stock exchanges. The North Sea bubble was eagerly worked into every portfolio, going forward.

Now that North Sea oil and gas are racing to oblivion, the scene has changed. Those far-out hopes of cheap oil are much further out, and across the cold and dusty plains of Central Asia, or many thousands of feet below the Atlantic Ocean off West Africa. Through late 2002, and more strongly after the summer of 2003, the old-fashioned glamor of gold regained its glitter, reminding bourse players that times have really changed: no longer was gold some has-been plaything for eccentric specialists. Such real resources as gold, oil, metals and agrocommodities will necessarily feature in plays made by the investor community, going forward to a higher-energy-priced world.

Growth is good, but real growth has migrated – along with tens of millions of jobs – to China, India and elsewhere. With this real growth has come inevitable and ever-stronger oil and gas demand growth in Asia's new industrial countries. Rates vary by country, but are often in double-digit annual percentages, and demand potential is so close to open-ended or unlimited that back-of-envelope calculations soon throw up world supply shortfall as their bottom line. No official alarm bells rang in the high-energy West about this through the first two years of the millennium, but by 2003 this had changed. Asia's oil and gas demand can only grow exponentially, like Chinese, Indian, Pakistani, and Iranian car fleet numbers. How this coming crisis will play out is open for conjecture; but any hope that Asia's growing economies will suddenly switch to the New Economy model of near-zero growth and plenty of talk about "sustainability" as a nice surrogate for the real thing, is doomed to contradiction. Doubling times for oil and gas demand are about six or seven years for both China and India, and this particular countdown can almost be followed in real time. The end result, we can be sure, will in no way be good for growth, since it will lead either to open conflict for remaining oil resources, after triggering vast price rises for oil and gas, or to countries quitting the global economy and seeking autarchy. For world stock exchanges, too, the message is beginning to flash on the big boards of the 2005–08 horizon: real resources and gold are among the few future winners in a complete inversion of the plays that made money in the paper booms of the last 15 years.

The big loser, apart from growth, will be globalization. The term itself is but a new slogan for the early nineteenth-century idea of "comparative advantage," and is intrinsically and totally predicated on, and enabled by, cheap energy. From even the time of David Ricardo and his key fantasy of *classical comparative advantage* – by which banana producers do not build airplanes, while computer manufacturers do not produce coffee – globalization was the conceptual base for mercantile fleets and world trading in anything and everything. Basically, this requires energy – preferably the cheapest possible. Without cheap energy, regional and local autonomy, or autarchy are just as logical or reasonable, and autarchy precedes modern – that is, fossil energy-based – land use and economic organization. The simple reason for this is that local and regional self-sufficiency are more energy efficient. In a permanent energy crisis of the type heralded by Peak Oil, we shift totally away from every scrap of logic underpinning globalization, which has done nothing more than dust off and re-institute very classical North/South metropole-and-colony relations. These, the first time around, ended with liberation wars of the Malaysian Insurgency and French/US Vietnam war type. The bottom line for globalization is simple: without cheap energy it is as dead as a dodo, whether the bird is imagined as full-grown or still having what analysts call "upside potential," while forgetting that it is, sadly, extinct.

14
The Myth of Decoupling

Andrew McKillop

At present, as for the past half-decade, the terms *decoupling* or *delinking* of oil from economic growth, or *economy dematerialization* are little used in economic and finance journalism, remaining focus of a very few learned footnotes in dusty economic journals. However, a glance at the finance columns of major newspapers or economic journals from the 1980s and early 1990s will throw up stacks of references to these magic terms, whose meanings shift through a spectrum from the world of real economics to mythmaking and fantasy economics.

These terms date from the early 1980s. Oil prices at the time, expressed in 2003 dollars, stood at about $80 per barrel, attaining a peak of about $103 for a few months in late 1979 and early 1980. A way to cut oil prices became the Holy Grail of every strategist and politician, from NATO advisers and planners to organizations representing speedboat racers or market gardeners. Pretty obviously, using basic economics, one way to cut the price would be a fall in demand for oil, because in theory, at least, demand falls as prices rise. This is a nice theory, but is not always true. With energy, it unfortunately ignores the fact of advanced industrial societies being totally energy dependent, and the plethora of crucial roles energy plays in economic activity.

The quest for cheap oil began in earnest in the early 1980s, and energy facets of the economy achieved real prominence at that time. One example of this was the Washington-based Institute for Energy Economics, of which I was an early member. The Institute, originally called "Institute of Energy Economists," before the subtle name change, was founded by a coterie of Ivy League economists. Long before Enron came and went, they lobbied for the idea that energy trading could and would bring down prices. Today, such academic lobbies acting for cheap energy and oil have more than somewhat faded from the scene, and energy trading has lived its full 20-year life cycle, with a peak of political acceptance in the early to mid-1990s.

It is now associated with the Enron debacle, price gouging of customers, and power blackouts and brownouts. So-called energy traders claim they can "stabilize or reduce" energy demand *per unit of economic output* – but whether this is true or not, the slightest skim through any national economic or industrial statistics reveals only that energy demand *goes on growing*. World oil demand since 1983, for example, has grown by almost exactly 50 per cent. We have near-total dependence on oil and natural gas for food, plastics, fertilizers, pharmaceuticals, mining the metals for our transport equipment, and running our city skyscrapers – in fact for everything we call "advanced industrial."

ECONOMIC MELTDOWN AND NEW ECONOMY SLOGANS IN THE EARLY 1980s

In the early 1980s, following the arrival of the neoliberal, rhetoric-spouting Thatcher–Reagan duo, accompanied by the second oil shock due to the mullahs in Tehran, who cut off about 6 per cent of world oil supplies for about six months, there was a near meltdown of confidence on world stock exchanges. The ensuing economic slump was closely comparable to that of the 1930s. Unemployment, for example, went into double digits; many "smokestack" industries seemed to go up in smoke – whereas in fact they relocated and reappeared, even hungrier for energy, in China, Korea, Brazil, India and elsewhere in the developing South. Naturally and inevitably, energy consumption fell for a few years in the North, and either flattened or fell slightly elsewhere. All and any falls in oil and energy consumption were concentrated in the period 1980–83. This was no surprise, given that this recession was the severest since the 1929–36 period, and that its *real* cause was sky-high interest rates, which were at their peak in 1980–83. However, due to the political impact of the second oil shock – Tehran mullahs bringing Islamic fundamentalism to the headlines – oil and energy prices remained high, fattening the profits of US and European oil majors and giving economics professors their big chance of being acclaimed, or simply published and listened to. At that time of fear, tension and difficulty, they had only to suggest that falling demand for oil and energy in the North through 1980–83 was *not* due to self-induced slump, but to some imagined "structural change in the economy." Their then-favored one-word terms for this were "decoupling" or "delinking." Other terms much used in that heady period of New Economics (which was rather far from new, being a rehash of 200-year-old nostrums), included "dematerialization." In 2003, the favored term is "demand destruction." Whatever the favorite buzzword, economics professors, soon followed by business gurus, could say outright that steel mills would give way to pizza parlors, iron ships to Internet messages, that employment would become *flexible* (not their own, of course!), and that this low-energy, cutting-edge economy would elevate all world citizens to Nirvana.

We shall examine a few facts here, but what counts is that energy "decoupling," for any length of time, is totally impossible without economic slump and mass unemployment. No technological fix can sustain consumption levels we call "advanced industrial," – for food, metals, plastics, drugs, concrete, transport, electric power, heating and air-conditioning, and all the rest – without *very* large quantities of fossil fuels being used as the basis for these processed products. While energy consumption growth can fall below the rate of economic growth as services expand, and the economy can demand smaller amounts of new steel, aluminum and other metals (replacing virgin metals with recycled metals from the enormous stock in circulation), it is almost impossible that energy growth can fall behind that of population. The fossil energy-based economy and the civilization it supports are simply not sustainable without gigantic throughputs of energy. Proof of this is the simple fact that world energy consumption, in 2002, is around 20 per cent above that for 1990 (see Figure IV.14.1).

ECONOMIC RECOVERY, GROWTH OF ENERGY CONSUMPTION

World energy consumption recovered through the late 1980s, then increased again through the 1990s. In the developing South, notably the newly industrializing and future industrial superpowers of China, India, Brazil, Pakistan, Turkey and Iran, energy consumption growth remained firmly locked onto economic growth throughout the period – that is it stayed "close coupled," a 1 per cent economic growth rate needing close to or more than 1 per cent growth of energy use. Almost inevitably, energy consumption growth in fast-industrializing countries is *led by oil and gas*, because these are the easiest energy sources to use.

From the mid-1980s and through the 1990s energy consumption recovered in the North, as the "policy shock" of sky-high interest rates was eased, and the "dynamizing" recipes (including

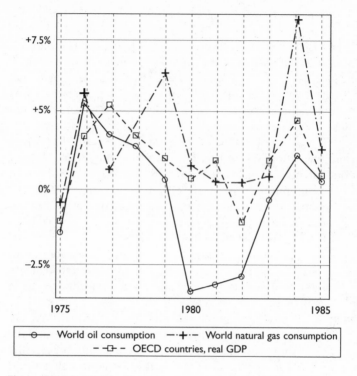

Figure IV.14.1 World oil and gas consumption and OECD GDP growth or contraction, 1975–85 (percentage change from preceding year).

the "discipline" brought by mass unemployment) of New Economics were quietly placed on the back-burner. At least until the Kyoto process became a subject of interest, and until Middle Eastern oil politics came back to haunt and excite politicians, war strategists and editorialists, the very idea of "decoupling" enjoyed no public significance, or airtime. Thus, as already noted, nobody talks about "decoupling," "delinking" or "dematerialization" at all these days; but this has already started to change. The 2003 variant is called "demand destruction" – a fall in energy demand because energy prices suddenly rise, instead of always falling as required to by New Economy diktat. Fervent appeals for "decoupling" will almost certainly come back to the surface, either through rushed attempts at finding some way to comply with Kyoto Treaty obligations (see

Chapter 10), or to ration and limit fuel consumption, when or if war and civil strife in the Middle East lead to a cessation of oil and gas supplies. For many countries with under-financed, under-maintained, and now undersized electric power generation and transmission capacities, electricity blackouts will pave the way for big price-hikes to ration consumption, if not to "destroy demand."

Figure IV.14.1 traces exactly what really happened through the oil shock and "delinking" period of 1973, before the first oil shock, up to 1988, when the Clinton Boom was being primed and readied for release on hungry equity markets, and after Reagan and the first George Bush had had their terms in the White House.

The actual pattern of the response of the energy economy to oil shock comes out clearly: the first oil shock of 1973–74 occurred at a time when political and business leaders in the oil-hungry industrial nations sought to maintain output and employment. At that time, the political call for "strong money" had not yet been pumped out in the media, and "interest rate medicine" had not yet been applied. This came with the Reagan–Thatcher duo, who seemed to have no comprehension that endless interest rate hikes themselves intensify *both* inflation *and* recession for quite a while. New Economics in the early 1980s destroyed tens of millions of jobs and tens of thousands of businesses, turning recession into a 1930s-style slump, and causing entire industries – both those wasteful and efficient in their energy use – simply to close down. This was entirely unlike the economic retrenchment and recovery after the first oil shock of 1973–74. After that 295 per cent hike in oil prices, leaders in all OECD countries sought to maintain economic activity and employment, and did not hit the interest rate panic button. As a direct consequence, the pattern and timing of economic recovery was very different: economic growth and oil consumption, and all other forms of energy demand, quickly recovered in 1974–75. This retrenchment and recovery took at most nine to 12 months, whereas the same process took as much as three years after the 1979–81 oil shock.

THE ROLE OF MONETARIST MEDICINE: DIMINISHING RETURNS

The second oil shock could be thought of as giving birth to the world as we know it today. Political and economic policy responses were totally different to those following the 1973–74 shock. By 1979–80 a new crop of political leaders, preaching New Economics and strong money, and cultivating their own personal image of courageousness, had attained power in the UK and US. Their radical and dynamic shock treatment of the economy cranked up an intense recession, simply through winding interest rates into the stratosphere. Today, intense and inflationary recession again beckons the rich-world democracies, if only because of the vast, untreatable budget and trade deficits of the US (the budget deficit being entirely due to the Bush administration), the fragile dollar, and surely imminent and substantial increases in the cost of oil and gas – and therefore of electricity. The "conditioned reflex" to recession through the 1990s, featuring endless cuts in interest rates, as embodied by the "cheap money medicine" of the US's Alan Greenspan, will disappear from the scene. We will almost certainly return to the hard monetarist reaction that the Iranian Revolution brought on in 1979–80. Now as then, financial and monetary decision-makers could well indulge in an orgy of rate hikes.

Their mentor could be Paul Volker, Treasury Secretary of the first Reagan administration, who in the early 1980s took US base interest rates to 22 per cent. Minimum loan rates in high street banks exceeded 25 per cent. This sure enough "saved the money" but also destroyed the manufacturing base of the US economy, the UK economy, de-industrialized many regions in Europe, and set the Asian Tigers (China, India and other emerging New Industrial countries) onto a track of breakneck expansion, and of course very rapid growth of oil and gas demand. The hollowed-out, shell economies of leading industrial nations that now import nearly all their industrial products from Asia are claimed to have been "downsized" rather than de-industrialized, partly for the good cause of oil and energy saving.

If we examine the actual patterns of energy-economy change with recession of the self-induced monetarist sort (sky-high interest rates to "defend the currency") it is particularly the "swing fuel," oil, which is turned off, first and foremost. In addition, we find that diminishing returns soon start to act on the apparent process of "dematerialization" of the economy. Gas and electricity consumption generally fall much less and more slowly than oil demand – that is, they react much more slowly to recession. The overall pattern of falling energy intensity within the economy then begins to dissolve as the recovery begins. During the recession of the 1980s, the New Economy began to reveal itself as the kill-or-cure medicine that it is. By the end of 1982, falling energy consumption in nearly all advanced industrial countries demonstrated the diminishing returns achieved by application of these techniques as annual falls in oil consumption for the OECD countries began to decrease. By late 1982 and early 1983, energy demand growth – now led out of recession *by oil* – began to strengthen. Those economics professors who had proclaimed "delinking" and "decoupling" had therefore to fall back on highlighting certain quarterly periods, in certain countries where there were continued *falls* in oil consumption, but some weak *growth* of the economy. This was brandished as proof that decoupling or delinking had really arrived, and could, God willing, be made to continue. By 1983, however, the game was over, and everywhere that economic growth happened, oil and gas demand increased.

SELLING THE GOOD NEWS OF RECESSION

The media consultants and spin doctors close to the seats of power, those craftsmen working up and polishing diamond-bright two-minute landmark speeches for Thatcher and Reagan, found a rich lode in delinking and decoupling. Their speeches would laud this modernization as having *nothing at all* to do with economic rout, but everything to do with the arrival of computer-based industries and financial services which, along with take-out pizza parlors, needed "almost no energy at all." Coupled with the bourse casino, where silicon chips replaced paper currency, producing "weightless wealth creation," a heady cocktail of illusion was cranked up to explain away mass unemployment and increasing poverty, for the simple majority of persons and on a world level, and stratospheric salaries and perks for the smiling captains of the financial services industry.

Falling demand for almost any primary product you choose, from oil and bananas to copper and soybeans, was missed by labeling them "sunset commodities," whose prices could easily fall to almost nothing. The term "sunset commodity" (which we no longer hear applied to oil in 2003) was used to describe anything not needed by New Economy service workers, as "old stuff." Extreme falls in prices of commodities exported by many African and some South American and Asian countries brought ruin, and helped trigger civil and international war – but for the rich democracies, these sunset commodities, at fire-sale prices, proved a useful subsidy in the fight against inflation. With real resources (and later energy) getting

.cheaper each year, inflation could be concentrated in such socially approved activities as housing and property speculation for the masses – the flimsy base for the "feel-good" sensation of personal wealth animating most voters and consumers in both the US and several EU countries by the 1990–2000 period.

From the early 1980s, and especially in the US and UK, sky-high, incompressible trade deficits became completely normal, despite the real menace of monetary collapse whenever the world loses confidence in the dollar, or the British pound (before it is shifted into the protective cocoon of the euro). As today, US trade deficits in the 1980s were magicked away by a Reagan-like imperial shrug of the shoulders, by massive government borrowing, and by US stock exchanges pulling in speculative capital flows from around the world. Today, however, the US is learning what it means to have one of the highest world per capita national debts (around US$30,000 per head, or a total of US$6,190 billion in June 2002) and a currency that under such strained conditions may lose its value rather quickly. This context of "structural trade deficits" dates from the 1980s, and results at least partly from a foolish quest for delinking of oil from economic growth. In the very short-term future (2003–05) all this will be revealed by the desperate measures the US will be forced to take to prevent the collapse of its formerly rock-solid dollar – and not to "defend strong money," as intoned by politicians of the 1980s seeking to crank up the value of their currency. The future downsized version of that policy will be to try to prevent the *collapse* of the almighty dollar. In any scenario – election-year or not – US interest rates will soon be hiked along with those of Europe, exactly as was done in the second oil shock period; yet the real oil price today, in 2003, is about *one-half* of its inflation-corrected 1983 price.

Artful or willful ignorance of the fact of economic slump in the early 1980s was obligatory in order to give credibility to the notion of decoupling. During the 1980–83 period, interest rates hit their highest levels of the entire twentieth century. One question that no kudos-hungry economist of the time ever asked was: Why did economic growth recover at all? Around late 1981 it was more than forgivable to imagine that the US, European and Japanese economies were on track to descend into a long slump of the 1930–36 type – an outcome that was only turned around by spending for the war that this classic liberal recession made almost inevitable. For an idea of just how intense the worldwide recession of 1980–83 was, Figure IV.14.2 shows how economic growth, consumer spending

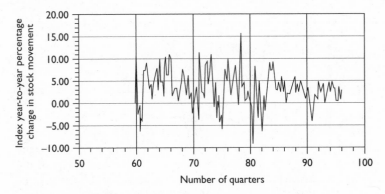

Figure IV.14.2 World stock exchange movements 1959–99 (Drexel Lambert website – no longer active).

and business confidence fell through the floor from late 1980, and stayed down for nearly three years.

NOT SUPPLY SHORTAGE, BUT PRICE RISES

In fact, the Armageddon and Economic Meltdown scenarios so deftly cobbled together in the early 1980s for the future of the high-energy-throughput economies – unless courageous medicine was applied to create mass unemployment – willfully confused an *imaginary threat to supply* of cheap energy with the *increasing price* of that energy. In the early 1980s only visionaries and eccentrics were allowed to talk, to themselves, about Peak Oil or sea-level rises through fossil-fuel burning. No real supply threat existed at the time. Actual supply cuts or shortfalls, due to Khomeini whipping up the chanting crowds in Tehran, had at most been about 6 per cent in volume terms, not much above or the same as the 4.5 to 6 per cent that had been denied to consumers at the height of the 1973–74 oil shock.[1] Even the most cursory look at Figure IV.14.2, or at energy and world economic output figures (Figure IV.14.3), shows that economic growth and energy consumption always came back. This is for what are called "structural reasons." More simply, an energy and oil-intensive, urban–industrial civilization with mass car ownership cannot destroy demand very much for very long without destroying itself. This will again be the pattern – until Peak Oil, that is. Perhaps ironically, as we shall note later on, world economic recovery itself is *intensified* by higher oil and primary product prices.

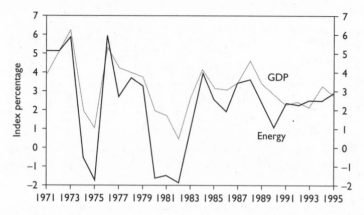

Figure IV.14.3   Percentage increase/decrease in GDP and world energy consumption, 1971–95 (Total-Fina-Elf).

RELINKED OR COUPLED ECONOMIC GROWTH
RIGHT THROUGH THE 1990s

When we check out the 1990s, long after any "decoupling" had disappeared from the scene by the simple fact of economic recovery and sharp declines of unemployment in the advanced industrial North, we find that world commercial energy consumption, through the Clinton Boom of the 1990s, was clearly linked, or coupled, as this data from the US Energy Information Administration shows. Yet another reason for this, of course, was that increased oil, gas and even coal consumption were to be logically expected following the oil price collapse of 1985–86. This "countershock" divided oil prices by three, and was later reinforced by the war reparations – in oil – that the international community was able to extract from Iraq, after Kuwait's liberation in 1991. The surfeit of cheap oil and energy available for roughly the period 1985–99 buoyed up and speeded the "recoupling" of energy with economic growth. Underlying the stock market boom of the Roaring '90s, therefore, was a quiet, constant increase in fossil energy consumption (see Table IV.14.1) in which the growth of oil consumption, at about 15 per cent, was somewhat higher than growth in total energy consumption (about 13.5 per cent). The so-called "dawn of renewables" is shown for what it is by the desultory 0.7 per cent of world energy contributed by renewables in 2000, even if their growth was

Table IV.14.1 World primary energy production 1991–2000, quadrillion (Peta) BTUs

Source year	Petroleum	Natural gas	Coal	Hydro electricity	Nuclear electricity	Non-hydro renewables	Total Primary Energy
1991	135.90	76.80	89.70	22.99	21.29	1.82	350.63
1992	136.50	76.90	90.20	22.94	21.36	2.02	352.28
1993	136.53	78.41	87.74	24.30	22.07	2.11	353.46
1994	138.31	79.17	89.39	24.47	22.50	2.22	358.49
1995	141.48	80.26	91.84	25.71	23.31	2.28	367.44
1996	144.95	84.01	92.60	26.10	24.13	2.38	376.79
1997	149.02	83.95	95.78	26.74	23.90	2.50	384.42
1998	151.90	85.65	93.97	26.65	24.42	2.61	387.71
1999	149.68	87.57	92.66	27.08	25.21	2.85	387.73
2000	155.25	90.83	92.66	27.52	25.66	2.99	397.49

World oil demand 1990: 66.74 million barrels/day. 2000 demand: 76.75 million barrels/day.

Quadrillion BTU = 10^{15} (Peta) BTUs (Note: Above data excludes resource energy input to Total Primary Energy. Growth of total energy input to provide TPE output through 1991–2000 likely exceeded 20 per cent.)

1 Quad BTUs = 293 billion kWh = 184 million barrels oil (equivalent).

Source: Energy Information Administration, USA, 2001.

impressive. In short, renewables in the 1990s had increased from nothing at all to not very much.

THE FINAL – AND REAL – DECOUPLING TO COME

The bottom line in all this is that within at the very most 35 years, but more likely 20–25 years, some rather hefty "delinking" will occur. By, at the latest, 2025 there will necessarily be a cut of over 50 per cent in the use of fossil fuels, extending by 2035 to more than 75 per cent, relative to today's levels, which for world oil consumption is about 78 million barrels/day. Prospects for achieving the extraction and production of about 115 million barrels/day by 2020, published as a supposedly feasible target by the International Energy Agency,[2] with similar figures from Exxon-Mobil and Chevron-Texaco – but not BP-Amoco – are in the realm of pure fantasy. The rate of depletion of oil is much faster than for gas because we are so close to Peak Oil. Exxon-Mobil estimates the overall economic + geological depletion rate at some 3.25 million barrels/day lost, on average, each year, for the period 2003–15. This is *cheaply producible* oil being lost, and replaced slowly and expensively by increasing amounts of what is called *unconventional* oil, which includes deep offshore oil (now below 8,000 feet), and synthetic oil such as Albertan "syncrude" from oil shales. This is essentially mined, using vast amounts of natural gas to soften and convert an almost solid substance in an environment that has daytime winter temperatures around −35°C. Currently, about 11 million barrels/day out of the total 78 million barrels/day the world uses in 2003 is unconventional. A very large part of the hoped-for increase in total world output will have to come from unconventional oil, and this alone is a major reason for doubting that world supply can exceed 90 million barrels/day for any length of time, and perhaps never achieve it.

By about 2020, whatever economic activity then exists – necessarily different from what we know today – will be decoupled. Conversely, in the 1970s, 1980s, and 1990s there was no decoupling of oil and other fossil fuels from economic growth. Whenever economic growth occurred, fossil energy burn increased. The only factors affecting the time taken for recoupling were the depth and severity of economic recession – either of spontaneous and bourse-led panic, in the first oil shock adjustment period, or deliberate economic slump through gouging interest rates, in the second oil shock adjustment period.

ELASTICITY – AND ELASTIC NOTIONS OF REALITY

Even the most cursory skim through any economics textbook will dig out the magic word "elasticity." Like "inflation" and many other economic terms, this everyday word is essentially not scientific, and has its meaning shaped by what the economist writing the textbook wants to prove. In economics, "elasticity" is usually a complicated way of saying that if something costs more you use less of it, for the same amount of satisfaction; or cut down the use of that more-expensive thing, and the satisfaction you get out of it, through "trading" how much more you have to spend, against how much satisfaction you lose. What counts for oil, gas and fossil energy (all of which are interdependent in terms of prices) is that oil price rises of 295 per cent in 1973–74, and 115 per cent in 1979–81 led to a fall in energy consumption, and that this fall induced, or reinforced, a decline in economic output *for a certain period*. The 230 per cent oil price rise of 1998–99, conversely, served to *further relink* oil demand with economic growth, for the simple reason that oil prices had fallen to such low levels.

Especially in 1979–80, with oil prices in today's money hitting about US$103 per barrel, it is no surprise that oil consumption fell;

but the first and second oil shocks had shocked more than just the comfortable business-as-usual gulping of non-renewable resources by the advanced industrial nations, wallowing in the cheap energy that all its citizens take for granted. Oil shock was interpreted by some, then by many, as denial of vital resources, a challenge to our civilization, and a mortal threat to the stability of the money in consumers' pockets.

ECONOMIC OBSCURANTISM

The econometric study for the 1974 "Project Independence" of the Nixon administration – promising "energy independence for the US by 1980 or at latest in 1985" – was carried out by the Harvard Business School professor Eric Zausner. His study, costing US$20 million in 1974 dollars (US$58.9 million in 2003 dollars), concluded that "price elastic" impacts of the oil price increasing from US$4 per barrel in 1973 to over US$9 per barrel in 1974 would reduce oil demand growth in the US to zero, and hold it there. Despite the oil price in 1975 then increasing to US$12 per barrel, and beyond, US oil demand *increased* by nearly 6 per cent in volume terms through 1974–76, notably because annual economic growth averaged nearly 4 per cent, on a real GDP basis, in 1975–78.

Curiously, higher oil prices were already at that time given the label "recessionary." This despite the simple fact that – after a single year of recession – US and other OECD country economic growth took off virtually as if nothing had happened. However, there had been a rise in inflation, and this was the focus of all the attention of the academic elite. To make their studies more appetizing and interesting to political decision-makers, the academics first claimed that "recession" was caused by high oil prices, and then transmuted this to "long-term inflation." Their approach has not changed. The New York professor Edward Renshaw, when setting out to prove that the early 1980s recession was almost single-handedly due to Jimmy Carter, had this to say in 2001 about the energy link with recession:

> Rapid increases in oil prices in conjunction with weak economic indicators are of particular concern since they may indicate that the economy has already slipped into another recession. The ten poorest growth years for the US economy from 1948–95 were all preceded by [an increase] in the price of domestic crude oil of 5 percent or more and a June–December increase in industrial production of 0.1 percent or less.

While higher oil prices may or may not be the [direct] cause of economic recessions they can lead to restrictive monetary policies that will help to terminate a business expansion. Except for 1953 (when the consumer price index was still increasing at a modest 0.7 percent annual rate) and 1971 and 1976 (when the U.S. economy was beginning to recover from economic recessions), and in 1996, the Federal Reserve prevented money supply M1 increasing when crude oil prices were increasing at an average rate of 5 percent or more ... the Fed has often resisted the inflationary effect of large increases in crude oil prices by [increasing] short-term interest rates ... at a rapid rate.

E. Renshaw, "Inflation and Natural Resource Scarcity: Can There Be Another Recession Without an Oil Price Shock?" (2001)

Like any true-grit economist, Renshaw sets out to confuse things, first saying that oil prices don't cause recession, then saying they impact on monetary policy, which is tightened, with this resulting in recession. However, we can give an answer to that agonizing question he poses, in the form of a very confident "Yes!" Recessions are very easy to crank up without a penny being added to the oil price. Worse still, for him and other economists, recession is one sure result of a collapse in oil prices.

THE REAL IMPACT OF OIL SHOCKS: WORLD ECONOMIC GROWTH

The real shock for fans of the now-forgotten slogan or doctrine of decoupling – which is somewhat shop-worn for the moment but will surely be dragged back into the limelight when oil prices regain say, two-thirds of their 1983 price levels in real terms (around US$60 per barrel in today's money) – is that, at least since 1979–81, and probably since 1973, oil price rises have *saved world economic growth*. There are several conditions and qualifications that apply to this heretical notion. The first is that this mainly applies to the consumption-saturated North, or advanced service economies, which are based on a huge, energy-intensive infrastructure base, and cannot possibly delink; that is, until their economies collapse. This may occur soon, or may even take as much as 15 years, but is not the point made here. The little-known, deliberately obscured fact is that oil price rises *restored* or re-dynamized conventional economic growth, especially through 1973–86.

After the 1985–86 oil countershock or price collapse, caused by deft market manipulation backed by Saudi connivance in oversupplying the oil market, and the later landing of some 400,000 troops to liberate Kuwait in 1991, no boom in economic growth occurred anywhere, when oil prices were restored to reasonable levels. With low oil prices the slow-growth Northern economy continued slowly growing, while simply counting down the interval before a classic liberal, or deflationary recession was triggered. Belief in the supposedly magical influence of cheap oil helped stock exchange operators push equity prices to wild extremes – triggering the 1987 world bourse crash, the worst since 1929. The hesitant recession that followed in 1989–91, and those "mild" recession-like intervals occurring in the 1990s, were arrested firstly through interest rate reductions. The paper recoveries were each time subsidized by cheap oil and gas, cheap sunset commodities, war reparations extracted from the crippled regime of Saddam Hussein in Iraq, and cheap capital pumped from the victims of "structural adjustment" in Africa and Latin America. This colonial tribute was then vastly expanded by the implosion of the USSR and its rapid conversion to Third World indebtedness and subservience.

From 2000 a full-blown and very classical recession came into being – despite interest rates at all-time lows. There was a "hump" or interval from about the first to the third quarter of 2003, when so-called recovery appeared to exist, especially in the US; but interest rate medicine will soon make that hump but a fleeting memory. One major reason for the post-2000 recession, other than the so-called dotcom and technology booms collapsing under the weight of their own hype, is that oil prices did *not* rise enough through 1999–2003. Conversely, in the first and second oil shocks, price rises were sufficient to re-start the motors of world economic growth. This happened rapidly after the first oil shock – in at most nine to 12 months. Economic recovery after the 1979–81 oil shock also occurred, but was much slower, being handicapped by sky-high interest rates. Conversely, the Clinton Boom faded and failed because it was almost exclusively internal to Northern economies, without any prop of higher economic demand *outside the North*. Just like any paper boom, it was condemned to fade away, then fail. Oil shocks provide one effective antidote to this.

WHY AND HOW OIL SHOCKS RESTORE GROWTH: CONSUMER SATURATION IN THE NORTH

This mechanism is very simple. Classic recessions, at least in the period since about 1950, are mainly due to *saturation of consumer demand* (which principally concerns the advanced industrial consumer nations of the North). After the pleasures of the first or even second family car, and especially that sophisticated pleasure obtained from trundling about in a two-ton 400hp, 4WD sport utility vehicle (often transporting just one happy consumer while spewing out very impressive levels of greenhouse and toxic gases), buying and finding the time to run a third or fourth family car is of declining utility. Even with the best, most civic greed in the civilized world, fewer consumers line up for their ignition keys. The same applies to the fourth or fifth family TV set or entertainment console. The same applies to mobile telephones, and especially when the crisply-printed bills flow in, or press articles hint your text messenger may give you brain cancer. The same applies to designer clothes, lifestyle cosmetics, plastic surgery and any other gadget or gimmick available. Of course, the huge industry of marketing and publicity sets out to smash all limits on individual greed and personal egoism, but a first victim of recession is falling advertising budgets which mirror weakening consumer demand.

INCREASED SOLVENT DEMAND IN THE CLOSE-COUPLED SOUTH

This does not take place in a vacuum. Some five-sixths of the world's population does not even have one motor vehicle per household, and 900 million people do not have enough food. Out there, oil price rises have an indirect *but fast* economic impact totally unlike the result predicted by well-fed politicians in the North who bleat: "These oil price rises will hurt the world's poor." The economies of the South are *close-coupled*. When and if oil prices rise, their own economic systems respond quickly, and rapidly increase output. Oil and energy price rises "trickle down" fast into rising prices for energy-intensive metals and minerals and agricultural commodities, and higher prices for the tourism services that countries of the South can charge Northern consumers. For the 1973–75 period estimates suggest this transfer of wealth from North to South was about 2 per cent of the North's GNP. Today, even when oil prices reach US$60 per barrel, such a transfer would barely scrape 0.75 per cent of the North's GNP. Price rises for raw materials, goods and services supplied by developing countries of the South, exactly like those inside the Northern economies, always include an inflation differential. That is, resulting price rises are greater than those due to solely increased energy prices. In other words, real revenue inflows to the

South *increase*. Since they are not consumer-saturated economies, this increase in solvent demand, and rising world liquidity – which also makes loans easier to obtain – leads to a rapid increase in consumption and imports – both from the Asian Tigers, and later also from the North. While this net increasing revenue flow to the South is often called "an unjust energy tax on the North" by the propaganda circuit of finance analysts, economists, journalists and politicians of the North, it in fact *restores* economic growth in their own economies by providing solvent external demand for those goods that cannot be sold to Northern consumers wallowing in a riot of consumption that periodically or cyclically slumps, due to saturation. That is: new solvent demand *trickles up*.

The new demand generated by oil price hikes and their downstream impacts on energy-intensive primary products first goes to countries and producers of lower-value industrial goods, from cassette players through home appliances and clothing goods, and then goes to higher-value and higher-technology consumer goods. All of these are supplied by the Newly Industrializing Countries – in the 1970s and 1980s the Asian Tigers, now also including China and India – whose economies make a rapid and price-adapted response to any upsurge in solvent world demand. Consequently, the Asian Tiger economies reacted fast to Northern recession and political agonizing in the second oil shock period. Table IV.14.2 illustrates this energy-economic response by three of the Tiger economies to nominal (pre-inflation) oil price rises of about 405 per cent, between 1973 and 1981.

This upward, near-unstoppable leap in oil consumption was *directly due* to the knock-on effects of oil price rises resulting in increased solvent demand by the developing countries of the South. In the simplest terms, the outward-oriented, growth-seeking

Table IV.14.2 Asian Tiger close-coupled adjustment to oil shock. Oil Consumption, thousand barrels/day

	1975	1976	1977	1978	1979	1980	1981	Increase 1975–81
Singapore	141	165	165	170	183	181	208	47.5%
South Korea	278	310	371	426	480	475	497	78.8%
Taiwan	214	271	304	353	358	388	359	67.8%

Source: "BP Statistical Review of World Energy."

economies of the Asian Tigers (and now China and India) are "close coupled." These economies receive the solvent demand generated by what is sometimes called "an unjust energy tax on the North" by opinion formers in the North, although all oil and energy consumers in the world pay it. This "pump priming" restores world economic growth, but requires a certain time to do so. The process operates as follows: "intermediate economies," or the Newly Industrializing Countries such as the Asian Tigers, grow rapidly through the receipt of new solvent demand that the North is temporarily unable to satisfy (for example, because of inflation inside the Northern economy making its export goods too expensive to compete with Asian Tiger exports). Due to this growth spurt, the Asian Tigers then seek investment and high-tech goods (called "capital goods," such as production plant equipment and machinery) and financial services, both of which are supplied by the North. At this stage – around 15–24 months after the beginning of oil shock – internal demand in the North is on the recovery track, after the essentially *psychological* shock caused by entry to recession. This can be expressed even more simply, by saying that a year or so of "fasting" by Northern consumers, overwhelmed by their riot of consumption (and told by their leaders this is a mark of cultural superiority), creates new demand in the North, as well as that already triggered and generated in the South by oil shock. Therefore, even in the late 1980s and early 1990s, even after New Economics and no-hope politics, the North *re-linked* or *re-coupled* economic growth to oil consumption.

UNJUST ENERGY TAX, OR ECONOMIC JUSTICE?

Economic recovery through *increased external demand* is ignored or denied by the entire decision-making and opinion-forming apparatus of the North. It may be cultural: the displeasure of knowing that brown people might have enough to eat and live slightly more comfortable lives, and not want to crowd into rusting boats, and dart through border crossings to serve as cheap labor and flesh in the North's shiny cities. It may be political: every single unit of extra economic output in the South means less need for developing country leaders to toe the Northern party line, to exert independence of judgment, and even say so. They may choose not to fall in line with whatever great struggle is underway (currently, the War on Terror) or being planned in the name of the New World Order. Whatever the reason, taking away a little poverty in the South produces an immediate loss of that so-comfortable superiority lurking in the unctuous spiel on "development imperatives" that feeds the huge economic development policy industry of Northern institutions, think-tanks and agencies.

Ideology apart, in the real world this mechanism has been triggered after each and every oil shock, including the price rises of 1998–99, and operated in the reverse direction after the countershock of 1986. While no sane politician or economic guru in the North would ever seek to put his comfortable job and personal prestige in jeopardy by saying so, the so-called Third World debt crisis, leading directly to the early deaths of millions of children in the 1986–95 period through increased poverty, was in part due to worldwide deflation triggered by the oil countershock. Since then, the Northern economy has been "retrenched," shuffling its stagnant wealth around and taking every step to prevent any "leakage." Oil and commodity price-hikes breach that firewall.

Through the unfettered working of global economic forces, oil shocks increase well-being in the South, increasing world economic growth and opposing the intrinsic deflationary trends of the consumption-saturated North. Using any economic or well-being index that any economist, politician or fit-to-print commentator might choose, the periods following the first and second oil shocks – of 1973–81, and the period following the countershock of 1986 (say 1986–95) – are closely compatable, for the two-thirds of the world's population living in what are called the developing countries of the South.

In the first period – when oil prices increased vastly – there was rapid and sustained improvement of living conditions in the South. In the second case, after 1986, when oil prices fell by around 65 per cent, and prices for all commodity exports of poorer countries took a nosedive, the South entered a long period of rising poverty, economic breakdown, civil strife and unrest, increasing malnutrition and increased child mortality. This was not only condoned but deliberately *increased* by what has been nicknamed Belsen Economics, a misguided and evil notion that the increased misery of the world's majority can somehow buy additional time for the "miracle of consumption" in the Northern economies.

CONCLUSION: DECOUPLING AND PEAK OIL

We should have no illusions about the intellectual or factual basis of "decoupling" or "delinking." The entire theory of "delinking" was and is as fragile, arcane and absurd as were the miracles of neoliberal economic boom and the degenerative cultural experience of post-1980s urban–industrial civilization that New Economics and No Hope politics imposed in their fevered, chaotic rush towards the lie of Universal Prosperity. The coming oil shocks, this time due to Peak Oil and depletion, will have a transient role of increasing incomes and economic growth to a real extent in the South, while of course *accelerating the depletion process* by raising world oil demand. With unfortunately little doubt, Northern leaders will first react to what will be final, or perpetual Oil Shock with Middle Eastern and Central Asian oil wars, and self-imposed economic slump through interest-rate hikes, because their ideology extends no further than these knee-jerk responses. It is ironic that in the very near term, over the period stretching to about 2005, any oil shock might, in fact, pull the Northern economy out of the current, classic liberal recession that threatens every day. The sooner the oil shock, the sooner will there be classic economic recovery. It will be a recovery like that of 1974–78, if interest rates are not sent through the roof. There will, however, be one very great difference, as noted throughout this book: we are entering the Final Energy Crisis. This time around, there will be *physical shortage* of oil, due to geological limits and not political embargoes. In economic terms the final result will inevitably be inflationary recession.

15
Crash and Crumble: Oil Shocks
and the Bourse

Andrew McKillop

All you need is $40 oil to bring the economy to a complete standstill. If we
have $80 oil we're going to be in the hole.
  Adam Sieminski, global oil strategist at Deutsche Bank.
                                              New York Times, November 13, 2002.

We only have four previous oil shocks with which to prove (or
disprove) that sudden oil price changes – three upward "shocks" and
one large downward from 1973 to 2002 – have a dramatic impact on
the stock exchange. As tables elsewhere in this book show (see
Chapter 14), periods when there have been fast changes in oil prices
have at least sometimes – in fact often – been associated with rapid
swings of stock exchange *sentiment* and index numbers. But this is a
very long way from any causal relation, from any kind of "scientific
statement" that if there is an X per cent rise in oil prices in Y months
then we get a Z per cent fall in Big Board (NYSE) index numbers, or
those of other stock price averages like the Nikkei-DJ, German Dax
or French CAC40. These can swing wild and free, and with no
discernible relation to the "real economy" for any number of other
trigger factors interpreted by the "trading and investor community"
as affecting their future strategies. Going back to the first and second
oil shocks of 1973–74 and 1979–81, a casual glance at the 50 per
cent or 60 per cent falls experienced by most leading stock exchanges over a few months could justify their being attributed to the
oil price factor. The fall in confidence and in index numbers could
also be explained, even more easily, as due to geopolitical uncertainty. That is, uncertainty and fears due to the 1973 Arab–Israel war
and Arab OPEC oil embargo, and the 1979 overthrow of Shah
Pahlavi, the "Gendarme of the Gulf," also called the "Second Eye of
the US in the Middle East," and considered along with Israel to be
the guarantor of the region's stability, through the all-seeing eyes
of his secret police. In US eyes, Shah Pahlavi assured constant supplies of oil at reasonable prices. Thus, there was surely an *oil linkage*

in the bourse panics of 1973–74 and 1979–81, but no single, oil-only cause.

THE PROBLEMS OF HYPOTHETICAL VALUE AND
THE *REAL CAUSE* OF CRASHES

There is plenty of ground for arguing that the 1987 stock market crash – in which worldwide losses of market capitalization were about US$1,850 billion, compared with about US$500 billion for the 1929 crash (both figures being in comparable-value dollars and also very approximate) – was in part due to *falling* oil prices. The mechanism is very simple: the big fall in oil prices through late 1985 and right through 1986, leading to prices being *divided by three*, was welcomed as a return to the happy days of the 1950s, because prices of around US$11.50/barrel in late 1986 were similar, in real terms, to prices of the 1950s and 1960s, before the first oil shock. In the 1950s cheap oil in the US was like a birthright, like running water – available as a universal or public good. The 1950s were economic good times, at least in theory, so a return to 1950s price levels for gasoline could only be good for business and good for stock markets, at least in theory. Consequently, share values and stock market indices were pumped up to delirious levels in late 1986–87, as unreasonable optimism spread through the investor community – unreasonable optimism spreading just as fast and furiously as unreasonable pessimism, we can note. Unfortunately, no spurt of economic growth happened anywhere in the OECD nations through 1986 and 1987. By October 1987 stock market expectations, that is, share values were far out of line with share dividends and companies' profit performance. So, there was a crash. Equally, anyone who wants to can argue that this crash had such little general economic impact (no more than six to nine months of slight recession in some countries) because bourse trading had already become *disconnected*, or unrelated to the real economy. Other arguments can be advanced – for example, that the 1987 Wall Street crash was palliated by continuing and vast US federal deficit spending, or even that cheap oil (following the 1986 countershock) might have helped "retrenchment" after the crash. Almost anything can be used to explain bourse crashes and their aftermath; ex post facto analysis and theory-building is a major activity in economic and financial circles. To prove this, any number of learned theories concerning the 1929 crash can be found, and most will cite different causes.

It has to be emphasized that stock market capitalization is a nominal or *hypothetical* cash value of all stocks, multiplied by their trading value on a given day. This is fine, but it would be totally impossible for all stocks to be sold on one day, producing or yielding the total market capitalization figure (for London's FTSE index, about £8,000 billion in late 2002). In fact, as every stock market "crisis" shows, whenever large numbers of players start to cash in their chips, just this fact alone impels a downward trend in average stock values, reducing the hypothetical Big Number.

NOTIONAL "VALUE" AND THE FICTIONAL PAPER ECONOMY

Stock market traders essentially trade notions of how the economy *might* turn out, sometime in the future. Operators need to draw in funds from the real economy, so they need to exaggerate earnings growth potentials a little, or even a lot. When "panic selling" is triggered for any reason, these notional future "values" are heavily depressed accordingly, but the "loss" of what never existed is hard to call a loss.

Consequently, there is a logic problem with the notion of "loss" on stock markets, because the total value of stocks in fact changes every day, and only a fraction of all quoted stocks are usually bought and sold on any one day. Any number of influences, including cycles and waves, play on trading sentiment and its outcome of stocks either "moving ahead" or "moving back." The degree of connection or linkage between stock exchange movements and what is called the "real economy," measured as barrels of oil consumed, airplanes produced or ordered, job creation and job losses, housing starts, text messages sent on mobile phones, services supplied, and so on, is at best tenuous. At times of bourse crashes the linkage can be *very* tenuous, because bourse players are reacting to political or military events, and talking up or talking down the index. Plenty of evidence supports the argument that stock exchange "disconnection" with the real economy only increases, and today ensures that stock market values are little related to the real economy.

More support for this argument comes from checking the amazing variety of pronouncements made by leading opinion formers, in their attempts to shore up artificial prices around the time of any stock market crisis. During the 1929 crash (see below) there were tragic, almost wistful, and increasingly desperate attempts to talk up unreal and overblown stock prices, in a context of constantly plunging index numbers, for about one whole year from November 1929. At the time, not a penny had been added to oil prices (they in fact fell, to around US$1.80 per barrel in 1929 dollars). In the run-up to the 1929 crash and for several years, however, it was not unusual for star stocks to put on 150 per cent in their notional value *each year*, with continual, reliable, week-by-week growth of the share price. As in the 1990s Clinton Boom, the reliability and predictability of these continuous rises in notional value itself led to unrealistic expectations by investors.

Anything can be grist for the mill of stock market sentiment. Since 1973, and especially in the last ten to 15 years, oil and energy price rises have earned a certain, usually supporting or secondary, but always *negative* role in bourse sentiment. Notably, through much of the late 1980s and nearly all of the 1990s, low oil prices ensured the subject was at best of marginal and sector-related interest, except during the brief spike of high prices just before the liberation of Kuwait in 1991. While oil price rises are described as negative sentiment shapers for stock exchange traders, they are *positive* for the earnings potential of oil and energy companies, whose shares are smartly bid up when oil prices rise. The standard mantra of any fact-oriented guru or analyst will be: higher energy prices "can only be bad for the bourse by taking money out of consumers' pockets." But some of that cash is put into the pockets of a happy few oil and energy companies, meriting a re-jigging of portfolios.[1]

THE 2000–02 CRASH

The agonizingly lengthy, slow-motion crash of 2000–02 is generally agreed to have cost about US$6,000 billion in lost or "now-you-see-it, now-you-don't" market value during winter 2002. The favored explanation of most fit-to-print finance analysts is that bourse players (but of course not they themselves) got carried away during the Clinton Boom. The very symbol of this classic boom in paper value was an explosion in dotcom, hi-tech and telecom share prices. Some brave retrospective attempts are made here and there to explain the crash by way of the 1999 oil shock, when oil prices increased from about nothing (around US$10 per barrel, or well below the 1973 price in real terms) to an inflation- and purchasing-power-adjusted price of around one-third of their value in 1981 (that is, to around US$30 per barrel in 2002 dollars). It could easily be argued that this rise in oil prices was and is *insufficient* to "prime the pump" of external demand – in economies outside the OECD countries, able by their increased demand first on the Newly Industrializing Countries, and then on the OECD countries, to counter the inherent deflationary and recessionary trends of the consumer-saturated, aging societies, and often stagnant economies, of the North. And in relation to the "dotcom bubble," most business observers have to acknowledge that an inevitable collapse of wondrously inflated overall share values riding on the dotcom and telecom boom was going to come, sooner or later. Given the revealed wisdom of US Federal Reserve Chairman Greenspan – that the dotcom and hi-tech economy produced vast increases in economic productivity – the oil price rises of 1999 should very easily have been accommodated, since, as noted above, oil prices only recouped about one-third of their 1981 value in real terms.

No fit-to-print guru or commentator cares to check out economic growth data and stock market trends the last time oil prices were anything like those threatened for the 2003–05 period. If they did, they would find that, far from wreaking inflationary havoc on investor confidence and squelching economic growth, much higher prices were in fact associated with belle epoque growth, and booming stock market expansion. Our guru will only have to make a few key-strokes on his personal digital assistant before the figures will appear on his screen: the all-time record year of economic growth in the US economy, *through the entire period of 1945–2002*, was 1984, with about 7.5 per cent growth of real GDP, year-round. Oil prices at that time, expressed in 2002 dollars, were about US$50–US$63 per barrel. Year-average stock market movements on the world's major bourses were up more than 20 per cent!

That was a long time ago – conveniently far back for our commentators and guns to have cobbled together and laid down the new wisdom that "high oil prices hurt economic growth." This myth, apart from explaining downturns in the now almost totally services-dominated, energy-lean economies of rich countries, also provides additional determination in fighting terror or changing regimes of various oil-producing countries. This of course is done only to provide greater liberty and well-being for survivors of the "therapeutic" bombing carried out, and will inevitably open up the country to strong investment potentials for overseas companies. This myth of oil prices hurting the economy, and their use to explain downturns in stock market index numbers, has also provided a handy reason for some commentators to explain the dotcom and telecom rout of 1999–2002.[2]

Speaking before a selected audience of sage economists and businesspeople, US Federal Reserve Governor Lawrence H. Meyer had this to say in June 2001:

> Why did growth [through 1999–2001] fall more sharply than anticipated and what does this tell us about the new economy? Sharp slowdowns are often the result of three inter-related and reinforcing developments: a coincidence of adverse shocks, an unwinding of pre-existing imbalances triggered by the deterioration in broader macroeconomic conditions, and a collapse in consumer and business confidence.
>
> The economy slowed in part, as I have noted, because monetary policy was committed to such an outcome. By mid-2000, it appeared that the economy, in response to the cumulative tightening over the previous year, was slowing ... The Fed stopped tightening and private forecasters were projecting a "soft landing."
>
> By October and November, it appeared that the slowdown was taking growth modestly below trend. Given the supply-side uncertainties I noted earlier, this outcome also seemed acceptable. But late in the year, the economy decelerated more sharply and we now know that growth fell to about 1 percent in the fourth and first quarters and it appears to have remained sluggish into the second quarter. We did not, however, have the data in hand at the time of the December meeting to confirm the degree or persistence of the slowdown. The Blue Chip consensus forecast in December, for example, still projected 3 percent growth over 2001.
>
> The sharper slowdown reflected, in part, the contribution of several additional shocks that reinforced the effect of the monetary policy tightening. Energy prices rose throughout 1999 and 2000. *Oil prices shot up in the fall and natural gas prices soared late last year* just as oil prices [had begun] to recede. The higher energy prices undermined consumers' purchasing power.
>
> Remarks by Governor Laurence H. Meyer before the New York Association for Business Economics and the Downtown Economists, New York, June 6, 2001. (Emphasis added.)

THE NEW ECONOMY AND GLOBALIZATION: DECLINING CONFIDENCE

The primary concern of Mr. Meyer, apart from trying to "talk up the index," was to repeat the considered and conventional wisdom, often pronounced by Federal Reserve Chairman Greenspan, that there is no alternative but the "new economy," despite its crushing and evident failure. And talking up the index is as necessary for the new economy as it is for the continued survival of the US economy's financial base – that is, sucking in speculative capital to play on US stock exchanges, providing an ever-necessary crutch for financing structural deficits of the US trade and finance accounts, and maintaining sufficient confidence in the US dollar not to expose the US economy to "imported inflation," through price rises due to a wilting greenback. While government deficits had been reined in through the Clinton Boom, they have now been brought back to life with a vengeance by George W. Bush's free-spending campaign against terror, and by large tax cuts being handed out, especially to the very highest earners. As Meyer went on to say in the rest of his speech (but not in these words), the threat of higher interest rates, when or if those speculative flows dry up, could lead to sharp, even uncontrolled decline of domestic US economic growth. The 2000–02 "correction" of US and other stock exchange numbers was in fact long overdue, given the sheer unreality of the dotcom and telecom booms. Yet Alan Greenspan still mumbles that these booms "prove" that productivity has risen by truly wondrous amounts, in the US and other advanced industrial economies. If this was the case, which it is not, there would be little problem for the US economy in absorbing and adjusting to oil prices equal in real terms to those of 1984!

The key element of *speculative capital flows* itself depends entirely on what is called "confidence" or "sentiment," and it is this *mass psychology* factor that is the one on which oil shocks act, like any other perceived threat to well-being and stability. Back in 1929, in the very old economy, perhaps the best known of all bourse crashes had definitively nothing at all to do with oil prices, but a lot to do with what makes stock exchanges tick and whirr. As economist John Kenneth Galbraith noted in his 1961 classic, *The Great Crash*,[3] stock market crashes, long before the "new economy," had at least the following in common with any and all crashes since the first oil shock: "[there is always] an atmosphere of unreality, gargantuan excess and menacing disaster" hovering over the scene, while the happy gamblers playing only to win "feel they have been providentially chosen to play in an unbeatable system," giving "riches without work." Little or nothing has changed since 1929 in what concerns the happy breed of investors placing their own, and above all other people's cash on various gambling chips chosen by whatever "split caps" operator (players' chips being split between supposedly high-risk and lower-risk plays) has the nicest line of spiel this week. Speaking before a UK House of Commons Treasury Committee hearing into the woeful implosion of "split caps" funds, the ex-Director of Aberdeen Asset Management, a UK split caps fund that had lost close to 90 per cent of its players' funds in twelve months through 2001–02, simply stated: "If markets do not recover, or if they fall further [then] the math is simple: they will not survive."[4]

While one or two of the elected representatives questioning this hero of paper trading accused him of being "the unacceptable face" of modern capitalism, his statement was entirely justified – if general index numbers on any bourse rise, and go on rising, then any fool (politely called "investor") can make money. If not, they lose. Stock exchange crashes, therefore, are entirely rational "corrections" when, for some reason or other, overall index growth has to be trimmed to reflect reality at least vaguely. This notably concerns the famous P/E or price–earnings ratio, measuring the cost of buying a share relative to the number of years that shareholding will take to pay off its acquisition cost and make a profit for the holder, through dividend and other payments that the holder receives (from profits and financial operations of the company or entity whose equity was purchased and held), of course with the risk that the company or entity does *not* turn a profit, or goes bankrupt. At any one time, the average P/E ratio for all shares listed on a particular stock exchange can be calculated. An intensive and purposeful reworking of various facts, figures and tidbits is used by the analyst community to do this, producing figures that range through a wide, even wild spectrum – in late 2002, for example, the index-wide or overall P/E ratio for the New York stock exchange, according to various gurus, was anything from 50:1 to 25:1. By late 2003, in the highly significant NASDAQ hi-tech marketplace, many enterprises, ever-hopeful of sucking in more capital, had real P/E ratios well above 200:1. If we take a P/E of 25–50:1 on a theoretical basis, the random purchase of any such listed stock would recover or recuperate the cost of its purchase in 25–50 years of holding that share. This is wildly too long, when we analyze the linkage of bourse plays in relation to the Final Energy Crisis.

COMING BOURSE CRASHES

The oil shocks that are within the next decade or two, aimed at the very physical base of our industrial civilization, will force overwhelming pressures for either total restructuring, or simple extinction. Any long-term placement not offering a P/E ratio of below ten or 15 at this time, late 2003, is simple suicide. As the above-cited hero of split caps rip-offs succinctly said, "they will not survive." But those fateful words apply not only to the cream of high-risk plays that finance sector "engineers" cobble together, but to any kind of placement in any enterprise or industry, outside a very small and focused range and concentrated in the sector of real resources. When oil and energy prices burst through a psychological barrier that we can place at around US$50 per barrel, the coterie of finance and bourse analysts and strategists will dust off and trot out their real resource-hedging and other defensive strategies. These will focus on oil and gold, energy, arms and defense industries, certain utilities, some agro-industries, government paper (such as Treasury bills), and of course hard cash. When oil prices continue to rise, as they will in the foreseeable future of the five-to-eight-year range, this defensive strategy will simply fall apart. As numerous chapters in this book show, there is no likelihood of cheap oil, or energy, surviving beyond about 2010–15, and the second date is generous: well before 2010 could be the "expiry date" for conventional or classic bourse plays. After the cut-off date we shall head towards the final bourse crisis, triggered and sealed by a combination of rising oil prices, geopolitical instability and oil war, climate change impacting agriculture with inevitably rising food prices, and exposure of the insurance industry to ever greater stress. The final bourse crash will be sealed also by ever-rising unemployment due to inherent deflationary trends and recurring economic slump, and the rapid aging of populations in the OECD North. By that time – in a period of much less than 15 years – the Final Energy Crisis will have fathered the final bourse crash.

THE 1929 CRASH

Before and during the 1929 Wall Street crash, the following are typical examples of attempts to maintain or restore "investor confidence" in the face of daily reality – and following the first round of the 2000–02 slow-motion crashes on world bourses, the same cheery optimism, misinformation and simple lies were again officially pumped out, as we shall see later:

> I cannot help but raise a dissenting voice to statements that we are living in a fool's paradise.
>
> E.H. Simmons, President, New York Stock Exchange,
> January 12, 1928.

There may be a recession in stock prices, but not anything in the nature of a crash.

> Irving Fisher, leading US economist, *New York Times*,
> September 5, 1929.

This crash is not going to have much effect on business.

> Arthur Reynolds, Chairman of Continental Illinois Bank of Chicago,
> October 24, 1929.

We feel that fundamentally Wall Street is sound, and that for people who can afford to pay for them outright, good stocks are cheap at these prices.

> Goodbody and Company market letter quoted in *New York Times*,
> Friday, October 25, 1929.

This is the time to buy stocks. This is the time to recall the words of the late J.P. Morgan ... that any man who is bearish on America will go broke. Within a few days there is likely to be a bear panic rather than a bull panic. Many of the low prices as a result of this hysterical selling are not likely to be reached again in many years.

> R.W. McNeel, market analyst, quoted in *New York Herald Tribune*,
> October 30, 1929.

Buying of sound, seasoned issues now will not be regretted.

> E.A. Pearce market letter quoted in *New York Herald Tribune*,
> October 30, 1929.

Hysteria has now disappeared from Wall Street.

> *The Times*, London,
> November 2, 1929.

The end of the decline of the Stock Market will probably be in a few more days at most.

> Irving Fisher, Professor of Economics at Yale University,
> November 14, 1929.

I see nothing in the present situation that is either menacing or warrants pessimism ... I have every confidence that there will be a revival of activity in the spring, and that during this coming year the country will make steady progress.

> Andrew W. Mellon, US Secretary of the Treasury
> December 31, 1929.

PLUS ÇA CHANGE

John Taylor, the George W. Bush Administration's international financial troubleshooter and a Federal Reserve advisor, in an interview with *The Times*, London, October 7, 2002 ("Bush Fireman Confident of Resurgence") had this to say: " ... we are going to have long expansions and short recessions. You have all the forces for recovery in place ... The upward momentum is there ... So I hope this expansion we are having now will be very, very long."

After the 1929 crash the American economy, and that of other industrial countries of the time, continued to decline up to 1933, and in some sectors and countries, to 1936. Typical falls in industrial investment and consumer spending were of 60 to 90 per cent relative to 1928. The rate of suicides in the US increased from a 1928 rate of about 95 to more than 150 suicides per million inhabitants per year through 1933–36, this rate being briefly approached (about 140 per million inhabitants per year) in the 1980–82 recession period, in both the US and UK.

16

The Chinese Car Bomb

Andrew McKillop

Until very recently, the exploding numbers of oil- and gas-fueled road vehicles – including cars, buses, trucks, motorcycles, scooters, mopeds, all-terrain "fun" vehicles and agricultural "off-road" vehicles increasingly used for road transport[1] – have drawn much less attention than human population numbers. This is curious given their rate of increase, shown by a few simple figures – in 1939 the world's roughly 2.3 billion inhabitants shared a total of around 47 million motor vehicles. Today's 6.3 billion human beings have around 775 million motor vehicles to fuel, repair, park and run, almost exclusively using petroleum and natural gas. Production of oil-fueled motor vehicles is increasing at least four times faster than human numbers in percentage terms. In several "latecomer" countries, vehicle manufacture is increasing the output and use of road vehicles at ten to 15 times these countries' rate of human population growth.[2] It should not be necessary to add that motor vehicles, a key part of consumer civilization, result in dramatic increases in personal consumption of oil and gas, probably in the range of 50–100 times, comparing "before-car" and "after-car" consumption habits.

Just as with the ultimate heat limit on world human population numbers (see Chapter 12), fixing a true and final limit on human numbers, there are set limits on the possible growth in numbers of motor vehicles. These notably include the ultimate reserve of petroleum, unless we wish to fantasize along with US Energy Secretary Spencer Abraham by giving any credence at all to his November 2002 statement that the world "will have a total of 3.5 billion motor vehicles by 2050." If this fantasy fleet were to come about – adding about 2.8 billion more vehicles to the world's stock – the fuel requirements for these vehicles, at current average consumption rates (see below) would *increase world oil consumption by about 70 per cent*. Because it is simply impossible to fuel this Fantasy Fleet on oil and gas, Abraham added that they would "of course run to a large extent" (not defined by Mr. Abraham) on hydrogen, the production of which he also of course did not explain.

To set an ultimate limit for petroleum and gas-fueled vehicles we can start with the near-ultimate example of a "latecomer" country in the car business – Japan. Even as late as 1949 Japan still had some 146,500 horse- and ox-drawn carts, compared with less than 200,000 trucks (and about 100,000 private automobiles). But through a self-reinforcing, mushrooming process of growth, with typical annual growth rates of 15 to 20 per cent, year in and year out, Japan's private car fleet explosively grew from these humble beginnings to its first million in 1963, to 5.2 million in 1968, and to some 26 million by 1982. Annual growth rates had by then considerably slowed, but the gargantuan size of the fleet itself allowed this slowed growth to give impressive annual increases: today's total number is about 45 million private vehicles.[3]

To get on the growth track, Japan's administrative elite, even after the culture shock of atomic weapons use against its civilian population and military rule by US Governor MacArthur, had to throw off mindsets dating from the 1918–39 interwar period, when road vehicles were seen as simple "feeders" for short-haul transport to rail, canal, river and coastal shipping points or transport nodes. Japan's domestic policymakers, at the start of the postwar period under US occupation, thus preferred to spend money on repairing and improving the rail, shipping and public transport sectors. In addition, their policy view downgrading road vehicles was reinforced by Japan's terrain, its dense urban centers, and by Japanese feelings of doubt on the safety of cars:

> Because of slow improvement of the country's narrow, often mountainous roads, the government tended to discriminate against motor transport on grounds of road safety. City streets were often dangerous too. There was strict traffic control, rigorous tests for driving licenses and careful inspection of all new vehicles, both home manufactured and imported. A high standard of maintenance was promoted and the manufacture of reliable, safe cars was encouraged.[4]

The date at which Japanese transport policy switched to outright support for cars can be set at about 1955–60, when animal-powered transport completely disappeared in the agriculture sector, together with the catch-up economic growth that Japan experienced from around 1958 – although as late as the early 1960s Japan's Economic

Planning Agency continued to underestimate the necessary forward growth of roads for the exploding numbers of vehicles.[5] Today, as the ongoing "restructuring" of Japan's national railway corporation proves – that is, the effective bailout of an underfinanced, neglected public transport system – public rail transport in Japan, as in its US role model, is a dwarf compared to the road vehicle sector, and national passenger transport depends almost entirely on the existence of private road vehicles.

While the US and, perhaps surprisingly, New Zealand had been the countries with the fastest-growing motorization in the entire period of 1905–40, achieving ownership rates for private cars of nearly 300 vehicles per 1,000 inhabitants starting from a near-zero base,[6] their growth rates peaked well before World War II; their "second wind" occurred – as for other leaders in the car owning pack, notably Canada, Australia, Italy, the UK, France and Germany – in the 1950–70 period. Typical growth rates included that of the UK, with its six-fold growth in car and private vehicle numbers through 1950–70.[7] At the time of the first oil shock of 1973–74, it was only Japan that experienced a strong (but short) downturn in this motorization trend. From no later than 1975–80 the tried-and-tested "economic growth model" of car-based and car-oriented growth – a key concept in economic mythology from the times of Henry Ford in the US of the 1920s – was applied with full force in several nontraditional car-owning democracies of the time, as well as dictatorships, including South Korea, Brazil, Malaysia, Turkey, Iran and the Soviet Asian Republics. Somewhat later (from the late 1980s) this growth strategy was adopted by China and India with no end in sight. Today, in countries such as Germany, the US, France and Australia, there is no difficulty finding three- and four-car households, nor 20-mile tailbacks every weekday on every main highway into congested, sprawling and polluted city and town centers. The same phenomena are to be found in São Paulo, Bangkok, Ankara, Seoul and Kuala Lumpur. In addition, a vast range of products arising from the magic of petroleum-based chemicals industries are essential to the modern private motor vehicle industry – notably plastics and resins. As in Henry Ford's time – when animal bone and ligaments, skins, wood and wood resins were still extensively used in car manufacture – the unfettered growth of the car industry remains highly attractive to economic planners, resulting in motorization continuing to spread out and away from the core countries of the aging advanced industrial OECD North.

There are, however, distinct limits on its ultimate reach. Today's private car and small vehicle ownership rate in the US is around 745 vehicles per 1,000 inhabitants, with lower but similar rates (around 500–650 vehicles per 1,000 population) in Belgium, Germany, France, the UK and other car-saturated economies. Applying the same ownership rate to India or China, and assuming these motor vehicles to be oil- or gas-propelled, results in absurd numbers for annual oil or oil-equivalent gas consumption. In the case of China's car fleet we are already, using World Bank data for 1990–99, at the fantastic but real average annual growth rate of about 18 per cent, doubling China's car population every four years.[8]

The following is of course a fantasy projection, but its only proviso is that India and China should firstly achieve at least the growth rates[9] for car production and ownership that were experienced by the US, New Zealand, Canada, Australia, Japan, France, Germany and other leading industrial nations, and then maintain these rates for a period of just under 20 years. If their growth rates are higher, the period needed to attain "saturation ownership" (the current US rate) will of course be shorter. If China and India were to do this, the entire oil exports of the OPEC group would not even satisfy these two countries' car fleet requirements for oil (or LPG/LNG)!

Average European vehicle mileage per year for the core six EU nations (EUR-6) is 22,000 kilometers per vehicle per year. The EUR-15 average is lower. Both average figures are rising with economic growth and the declining real cost of energy. Average vehicle occupancy is 1.5 persons. Annual mileage averages are not set to fall, nor occupancy figures to rise, except by decree or other draconian measures resulting from real oil shocks – for example oil prices above US$100/barrel. Average fleet-wide car fuel consumption in Germany is 7.9liters/100km per vehicle; but we will assume that this is rapidly reduced, for India and China, to two-thirds of that value, or 5.3liters/100km. We will also assume that Indian and Chinese cars will only travel 18,000 kilometers per year. Total oil consumption at 5.96 barrels (948 liters) per vehicle per year is calculated as follows. Using future population figures (assuming complete zero population growth) of 1 billion for India and 1.25 billion for China, we obtain a future car fleet – at the "saturation ownership" rate of 745 vehicles/1,000 population – of 745 million vehicles for India, and 932 million vehicles for China. At 5.96 barrels/year for each vehicle, their consumption is 5.54 billion barrels/year for China and 4.44 billion barrels/year for India, or a total of almost exactly 10 billion barrels/year, equivalent to 27.4 million barrels/day. This is about *three times total oil imports of all EU countries in 2002,* nearly three times the maximum possible production capacity of Saudi Arabia, and slightly more than the total average export volume of the OPEC group in mid-2003.

We therefore have a laughable fantasy: an insight into exactly why three nuclear-armed powers – China, India and the US – are ever more likely to fight among themselves, or confront EU importers, including two nuclear-weapons states for *the last oil reserves on the planet.* Under any hypothesis – excluding childish technological fantasies and utopias such as those trotted out by Amory Lovins or US Energy Secretary Abraham – there is simply no prospect of China, India – or other countries such as Malaysia, Brazil, Turkey, Iran, Ukraine, Mexico, the Czech Republic, and other emerging car producers – being able to achieve US, West European, Australian or Japanese rates of car ownership. The Chinese Car Bomb therefore ticks onward, as each day another estimated 112,190 cars are produced.[10] Each one requires up to 55 barrels of oil-equivalent to produce, and must operate on bitumen-based highways, on tires that themselves are about 40 per cent oil by weight. Not only is this explosion of the world car fleet a serious threat to the earth's environment, but through its oil demand impact it will become a threat to international peace and stability.

17
A Reply to "Global Petroleum Reserves – A View to the Future" (by Thomas S. Ahlbrandt and J. McCabe, US Geological Survey)

Colin J. Campbell

Ahlbrandt and McCabe have written an elegant article, choosing their words with extreme care, to present what seems to be an authoritative account of the world's oil and gas situation, based on a study made by the USGS in 2000. This is in fact a thoroughly flawed study that has done incalculable damage, misleading international agencies and governments, including even perhaps their military and strategic planners. The 2000 study was in fact a departure from earlier, sound evaluations by the USGS under its previous project director, the late C.H. Masters, who understood the situation well and used great skill in delivering the message, albeit at times between the lines, as he recognized its sensitive nature.

Neither of the authors claims practical oil experience, as is betrayed by their mindset, which is more appropriate to the mining geologist for whom resource concentration is as important as occurrence. They say they speak to a Mr. Green of Exxon, but we do not know what he tells them or why; another Exxon spokesman, when asked about the USGS study, reportedly made the succinct reply, "You get what you pay for, and that came free."

It is an old trick for the politician to provide an answer to a question that is not asked. No one need be seriously concerned about when the last drop of oil will be produced, when what matters – and matters greatly – is the date when the growth of production gives way to decline, because of resource constraints. This is the transcendental issue, given the world's dependence on abundant oil-based energy, which furnishes 40 per cent of all traded energy and 90 per cent of transport fuel, without which world trade will be crippled. The US itself attained peak oil production and shifted to decline in

1970, and the same pattern of growth to decline is being repeated from one country to another around the world – one of the most recent being the UK in 1999. Production inevitably has to mirror earlier discovery, after a time-lag. The world peak of discovery was in 1964, and it should surprise no one that a corresponding peak in production is now imminent. It is self-evident, however blinkered our eyes.

The authors present the comforting notion of the resource pyramid, implying that the world can seamlessly move to more difficult and expensive sources of oil and gas when the need arises. But there is a polarity about oil that they fail to grasp: it is either present in profitable abundance or not there at all, due ultimately to the fact that it is a fluid concentrated by nature, in a few places with the right geology. They speak of "crustal abundance" when a glance at the oil map shows clusters of oilfields separated by wide barren tracts.

They emphasize "reserve growth" as a new element, missed by their predecessors, yet fail to point out that the text of the study itself expresses grave reservations. "Growth" is in fact more an artifact of reporting practices than a technological or economic dynamic. In short, reserves described as *proved* for financial purposes refer to what has been confirmed so far by drilling, saying little about the full size of the field concerned. Clearly it was absurd to apply, as the study did, the experience of the old onshore fields of the US, with their special commercial, legal and reporting environ-ment, to the offshore or international spheres, where very different conditions obtain.

The authors speak of their impressive probabilistic methods, which in the study allowed them to quote estimates to three decimal places. In, for example, the famous case of little known northeast Greenland, the study states with a straight face that there is a 95 per cent subjective probability of more than zero – in other words, at least one barrel – and a 5 per cent probability of more than 111.815Gb. A *mean value* of 47.148Gb is then computed from this range, and is included in their global assessment. Can we really give much credence to the suggestion that this remote place, which has so far failed to attract the interest of the industry, holds almost as much as the North Sea, the largest new province to be found since World War II? Could this be pseudo-science?

Turning to the actual estimates, the authors state that the sum of past production, reserves, reserve growth and undiscovered fields comes to about 3 trillion barrels, but then claim that the peak of production will not arise before about 2050. Experience shows that the onset of decline comes at, or before, the midpoint of depletion, due largely to the immutable physics of the reservoir that impose a gradual decline in production towards exhaustion. Depletion midpoint on their estimates comes when 1,500 billion barrels have been produced, which will be reached around 2020 at present production rates, or sooner if demand should rise. A mid-century peak implies an utterly implausible precipitate fall. But even this line of reasoning does not itself paint the full picture, because it fails to distinguish the different categories of oil. There is clearly a huge difference between a Middle Eastern free-flowing well, and tar-sand oil resources in Canada mined with strip-mining shovels. As the authors themselves state, there will be an increasing reliance on heavy oils, low on their resource pyramid, which are slow to produce and will not contribute significantly until after global Peak Oil, for obvious commercial and environmental reasons. The USGS study did not itself forecast production, but simply indicated the amounts to be found over the 30-year study period. But the internal evidence, flawed as it is, speaks of a peak long before the mid-century. If that were not enough, we can now compare the actual results with USGS forecasts for the first seven years of the study period. The indicated average annual discovery is 24 billion barrels, whereas the actual has been *less than half* that amount. This is doubly damning, because it would be normal to expect the larger fields to be found first, as the past record amply confirms.

What the USGS failed to do was to extrapolate past discovery trends in the world's mature basins, containing most of its oil and gas, having properly backdated reserve revisions to the discovery of the respective fields. It is axiomatic that a field is found by the first successful borehole drilled into it, even if its size is not exactly known at the outset. Had the USGS done that, it would have had the benefit of the considerable experience of the oil industry working in the real world, which is likely to give a better view of the future than abstract geological assessment couched in subjective probability rankings.

The authors accuse those who draw attention to the manifest failure of the study of having hidden agendas, introducing the colorful but unhelpful designations of Cornucopian and Malthusian, when all we seek is a realistic assessment of this critical issue.

The article reviews two specific areas: the Caspian and Iraq. Is it a coincidence that the US earlier attacked Afghanistan, which borders the Caspian, and has since turned its guns on Iraq? Let us hope that its foreign policy is not being influenced by this thoroughly flawed work. While one can forgive its authors for having got it wrong, as it is a difficult subject, to persist with the error using persuasive language and specious argument verges on the culpable.

18
Price Signals and Global Energy Transition

Andrew McKillop

Most economic policy-makers subscribe firmly to the belief that cheap oil and energy underpin economic growth. Very large amounts of fossil energy are certainly vital for any modern economy, whether the OECD bloc's service-oriented economy, or the fast-industrializing economies of the Asian Tigers in the 1975–90 period, or China and India, and other highly populous industrializing countries today. The absence of any "alternate model" for economic development ensures that there is continued and strong, worldwide demand growth for fossil energy. Upward potential for personal consumption of fossil fuels is essentially unlimited in this context.

The role of oil and energy price rises in increasing or decreasing economic growth, changing the type of economic growth that takes place, and either increasing or decreasing oil and energy demand growth rates, is not well understood. However, depending on the policy and fiscal context, it can be stated that oil price rises to high levels (probably up to US$75 per barrel) almost certainly increase overall, global economic growth rates, and therefore increase oil and energy demand growth rates. Only extremely high oil and energy prices, or extremely high interest rates and very deflationary economic policies, can abort this process.

Since about 1994–96 world energy and oil demand growth rates have increased dramatically. This "demand shock" is due to a number of economic, energy-economic, social and technological factors, and in the absence of grave economic recession, higher demand growth rates are likely to continue. Current "trend growth rates" for world energy and world oil demand are about 2.25 per cent for oil and about 2.5 to 3 per cent for energy on an annual basis, with major regional variations.

The "cheap oil interval" of about 1986–99 was an anomaly from many perspectives, and for many reasons. One key reason is physical depletion, which is nevertheless rejected or ignored by most governments and institutions as a price-setting factor for oil and natural gas. In relation to oil, more important than physical deple- tion in the very short run (the next three to five years) is the question of available production capacity, producer-country stability, and pricing policy decisions of OPEC. After 2008, the world oil market may face a critical structural supply deficit. Before that period, demand growth and loss of capacity through accidents, strike action, natural disasters, OPEC export limitations, and civil war or sabotage in exporter countries, will likely produce major price spikes.

OIL PRICES AND ECONOMIC GROWTH

In the Reagan re-election year of 1984 the US economy attained its highest-ever post-war growth of real GDP, achieving what today would be the unthinkable annual rate of 7.5 per cent. At the time, in 2003 dollars corrected for inflation and purchasing power, the oil price range for daily traded volume crude oil was between US$52 and US$65 per barrel (see Table IV.18.5). Despite this simple fact of economic history, cheap oil is still regarded by uninformed opinion, as well as by most government agencies in charge of economic man- agement, as a passport to economic growth.

Oil prices as high as US$60/barrel would not harm the world economy today. This would almost certainly herald increased growth of the world economy within a few months. Conversely, the setting of extremely high interest rates would result in massive economic damage. There would be inevitable collapse of world stock markets, runaway "domino-effect" bankruptcy of many major finance-sector corporations, mass layoffs and unemployment, and grave problems in financing the structural trade deficits of the US and UK, in particular. The US, also facing an all-time record deficit in its public finances (at least US$455 billion in 2003) and around US$5 to US$6 billion per month in costs from its regime-changing activities in Iraq, would expose itself to the risk of runaway flight from the dollar, just as the interest rate weapon produced stock market and economic rout in its wake. The declining petromoney status of the British pound would be unlikely to shield the UK economy from the consequences of using the interest rate weapon as a blunt instrument of energy policy to force down oil demand. All European Union countries and also Japan, would face severe national budget financing difficulties, as tax revenues collapsed and spending to limit economic damage, including unemployment compensation and bailouts for large companies, spiraled upward as the crisis deepened. The financing of increased state spending through borrowing would then lock on to the upward spiral in interest rates, and itself intensify recession while maintaining inflationary pressures.[1]

Higher and much less volatile oil and energy prices underlie serious and committed energy conservation, transition to renewable energy and restructuring for a low-energy economy and society: these are the real long-term solutions to emerging supply difficulties – which will certainly raise prices. But energy transition is discarded or rejected as utopian and unworkable by political decision-makers. While claims are made that today's economy is "less oil dependent than in the 1970s,"[2] world oil consumption has risen by about 48 per cent, or 20 million barrels/day since 1983, and by about 17 per cent since 1990. Oil import dependence as a percentage of total consumption continues to rise in a large number of OECD economies, and unless demand is rapidly substituted somehow, oil imports will soon show very fast growth.[3] Unfortunately, the subject of oil prices is given benign neglect when they fall, and over-energetic propaganda treatment when they rise. Most economic policy-makers believe in a simple slogan: the lowest price is always the best.

In theory, the "price signal" of higher oil and energy prices must be present if a range of goals – stretching from reduced greenhouse gas emissions, through energy independence, to slowing the rate of fossil-energy resource depletion – are to be regarded seriously. If they are not taken seriously, this can easily explain the basic unpreparedness of large oil and gas consumer countries to accept higher and more stable oil prices. Any large interruption in supplies, of more than 5 per cent or so for under six months – or depletion-linked failure of world production capacity to match demand and demand growth – as in the past, creates an immediate crisis.

This leaves demand destruction as the only real response to any large rise in oil or gas prices, through economy destruction by the interest rate weapon. The last time this was done, in 1980–83, oil prices were reduced through the cutting of economic activity in general. Oil prices in today's money fell from US$100/barrel in late 1979 to around US$60/barrel in 1984, but the collateral economic and social damage was awesome. Unlike today, however, the OECD economies then started from a position of growth, with balanced budgets in many countries, including the US, in 1979–80. The world economy was able to swallow the bitter pill of sky-high interest rates without imploding into a sequence like that of 1929–31. There is no guarantee that this would be the case today – no soft landing is currently on offer.

HIGHER OIL PRICES TEND TO INCREASE WORLD ECONOMIC GROWTH

Higher oil prices operate first to stimulate the world economy, outside the OECD countries, and then lead to increased growth inside the OECD. This is through the revenue effect on oil-exporter countries, and subsequently on countries exporting metals, minerals and agrocommodities most of them with low incomes (with a per capita GNP below $400/year). Almost all such countries have a very high marginal propensity to consume. Any increase in revenues, due to the prices of their export products increasing in line with the oil price, is therefore very rapidly spent on purchasing manufactured goods and services of all kinds. In the 1973–81 period, in which oil price rises before inflation reached 405 per cent, the Newly Industrializing Countries (NICs) of that period – notably Taiwan, South Korea and Singapore – which we can call "traditional" NICs, experienced very large and rapid increases in demand for their exports. These three countries increased their oil imports in under eight years, through the 1973–81 period, and despite the 405 per cent price rise, by 55 per cent to over 80 per cent in volume terms (see Table IV.18.1).

The macroeconomic mechanism of higher revenues completely displacing any price-elastic impact from much higher oil prices, working between real resource exporters and the "traditional" NICs, quickly ratchets up world economic growth (the very simplest type

Table IV.18.1. Asian Tiger, close-coupled adjustment to oil shock – consumption in 1,000 barrels/day

	1975	1976	1977	1978	1979	1980	1981	*Increase 1971–81*
Singapore	141	165	165	170	183	181	208	47.5%
South Korea	278	310	371	426	480	475	497	78.8%
Taiwan ROC	214	271	304	353	358	388	359	67.8%

Source: BP Statistical Review of World Energy, various editions.

of Keynesianism, but at the global level), and is easily triggered by rising oil, energy and real resource prices. This flatly contradicts the propaganda of certain well-known institutions that higher oil prices "hurt poorer countries the most."[4] Higher revenue earnings for many low-income oil exporter countries may even prevent such countries from experiencing the conflicts leading to stoppages of exports. For the special cases of Iraq and Saudi Arabia, higher revenues may be the only effective, short-term way to prevent complete chaos in Iraq, and for Saudi Arabia to avoid civil strife, insurrection and takeover by hard-line Islamists.

No immediate recession can occur with oil at US$50 or US$60 per barrel. Vastly higher oil prices than that would be needed to abort the worldwide mechanism of higher oil, energy and real resource prices driving faster economic growth. Conversely, low oil and energy prices producing low real resources prices, combined with rising population numbers, surely aggravate the "cycle of poverty" in low-income commodity-exporter countries. Deprived of sufficient revenues, such countries have become indebted "basket-case" countries, subject to draconian conditions from the Club of Paris, World Bank and IMF for debt refinancing and restructuring. Constant ethnic and civil war in Africa provides the most vivid example of what happens to countries subjected to so-called "structural adjustment." (See Chapter 6.) When or if this affects oil-exporter countries there can be no surprise if this reduces or eliminates exports by them which, after the "price-taker" stage, fall into the bottomless pit of basket-case, low-performing economies. When they fall from that into civil and ethnic war their capacity to supply oil – whether cheap or not – must also suffer.

Today's "emerging" NICs include China, India, Pakistan and Brazil. All have either big or immense domestic markets, and large potentials for military Keynesian spending – that is, safeguarding national economic growth through deficit-financed, labor-intensive modernization and expansion of their military systems. The relative lack of integration of these behemoth economies into the world system – particularly India and Pakistan – also affords them some shelter from the effects of world recession, when or if the OECD countries tilt towards all-out recession. Conversely, whenever any increase in world solvent demand for manufactured goods occurs, these countries rapidly increase output. China is now without question the world's leading industrial power for medium- and low-value consumer manufactured goods, and will soon become the world's largest industrial economy. Under almost any hypothesis, therefore, fossil energy demand – particularly for oil and natural gas – will increase in China and India, and in the other large-population NICs. Demand growth can only run at rates at least close to, or usually well above, their rate of economic growth.

WORLD OIL DEMAND CHANGE UNDER REGIMES OF RISING PRICES

Oil remains the economic "swing fuel" par excellence, and oil price increases – before reaching certain supposedly "extreme" levels – will always tend to increase or restore economic growth at the world level. Furthermore, oil shock or sudden large price increases, as well as slower-acting large price rises that do not fall back, change the *type* of growth towards more energy-intense industrial and manu-factured products, and away from more services-based, lower-energy activities.[5] This perverse factor results in the increased oil intensity of world economic output and raises the "oil coefficient," or per-centage increase in oil demand per percentage-point growth of the economy.[6] This macroeconomic change affects all economies, but some faster than others. Unlike the stock of myths and "facts" with-out foundation that circulate inside the oil trading community, these effects can be *measured* and have predictive value.[7] Briefly, a regime of higher oil and energy prices tends to ratchet up global economic growth rates. This, in turn, produces the perverse result of firm demand for much more costly oil and gas. Whether this is infla-tionary or not will depend not only on how high oil prices rise, but more importantly on the fiscal and policy environment in large consumer and importer economies.

WORLD OIL DEMAND POTENTIAL AND "DEMOGRAPHIC" DEMAND

So far as potential demand is concerned, any oil supplier (OPEC or not) should take heed, when serious analysis is given to real-world oil demand structures and growth-drivers. These are all, finally, due to demographic and economic growth, to conventional technology used in the economic process, and to the very slow progress in find-ing real, economic, and effective substitutes for oil, gas or even coal that actually deliver more net energy than they cost to produce. In addition, such is the utility of fossil-based liquid hydrocarbon fuels

Table IV.18.2   Demographic rate of oil demand, 2002

Country/Region	bpy	World demand at this rate
USA	25.6	445Mbd
Italy	12.4	215Mbd
China	1.45	25Mbd
Rural areas, LDCs	0.2	3.45Mbd
Real world	4.51	78Mbd
World annual population growth 85 million		Annual 'latent demand' increase 1.06Mbd

Sources: Population data from UN Population Information Network; oil demand, BP Amoco Statistical Review of World Energy, 2003.

and pipeline gas that sought-after substitutes must be of a type that can be used without total restructuring of either the economy or society.

Current oil demand worldwide extends downwards from 25.6 barrels/capita/year (bpy) for the US, to well below 0.2bpy in the rural areas of low-income developing countries (LDCs). The world average, which fell slowly for around 15 years through 1978–93, is about 4.51bpy. As a pure projection, if the world's current 6.3 billion population consumed oil at current US per capita rates, this would generate a demand of around 445 million barrels/day. At the other extreme, at 0.2bpy world total oil demand would be telescoped to less than 3.5Mbd. (See Table IV.18.2.) The current, real-world average of 4.51bpy is around one-third the average for European Union countries, more than four times that of India, and over three times that of China – which will soon become the world's biggest industrial economy. Annual increase of world population (which is continuing to fall both as a percentage and in absolute numbers) is now running at about 85 million. At the current world average of 4.51bpy, this itself generates a "latent" or potential growth in world oil demand of about 1.06Mbd annually, assuming no change in the energy economy, no fuel substitution, and no economic growth.

The following points are highly significant:

1. If world average oil demand per capita in 2003 was the same as in 1979 (about 5.53bpy with oil prices, in today's money, at up to US$100/barrel), world oil demand today would be at least *17Mbd higher* than it is. World oil demand in 2003 would run at an

average of about 95.4Mbd. There is no certainty at all that world supply could satisfy this demand.

2. If we take current "demographic demand" (4.51bpy), the growth of that demand due to population increase, of about 1.06Mbd per year, is probably an absolute minimum, except in the event of very severe global economic recession, with actual contraction of world oil demand.

3. Any sustained growth in the world economy – that is, recovery from recession in the OECD bloc – and/or continued fast economic growth in China, India, Brazil, Pakistan, Iran, Turkey and other highly populous "emerging" NICs, will significantly increase total annual world oil demand growth to far above 1.06Mbd, perhaps to its double. This latter is 2.7 per cent of 78Mbd.

4. Given that world oil demand has increased by about 12Mbd since 1991, and that "demographic demand" is slowly growing again, it is wholly unrealistic to imagine that cumulative world demand growth will be any *less* than about 12.75Mbd in the next twelve years. This would only change in the event of long-term and worldwide economic recession, or coordinated and legally binding world action for energy transition.

DEMAND SHOCK

The BP Amoco *Statistical Review of World Energy*, 2003 edition, notes what it calls "surprising growth" of world energy demand since 2001 – about 2.6 per cent per annum, compared with a so-called "ten-year trend rate" of 1.4 per cent for world energy, and 1.3 per cent for world oil demand by volume. These ten-year trend rates are also used by many energy companies and institutions, such as the US EIA and OECD IEA. However, such long-term trend rates of demand growth for oil, gas and coal were in reality already giving way to higher yearly growth rates by about 1995. It is difficult, or even impossible, to identify any price-elastic factor in these major changes, and the large increase in annual oil demand growth rates since about 1995 can only be understood as resulting from the revenue effect far outweighing the price-elastic effect, in global macroeconomic terms. It can be noted that the US EIA and OECD IEA, since 2000, generally refer to a trend rate of world oil demand growth in the range of 1.7 to 1.8 per cent per annum.[8]

By comparison, when oil prices are considerably higher than today's current levels, demand growth rates also tend to be higher. During the 1975–79 period, with oil prices in today's money in the US$38–US$55/barrel range, world oil demand growth easily averaged 4 per cent per annum by volume, after a sharp, one-year fall in 1975. This can be compared to the 1999–2003 sequence of world oil demand change, with a sharp fall in the single year of 2001. The fall in demand for 2001 against the previous year (about 1.2 per cent) could be claimed as a price-elastic response to tripled prices, around two years after the 1998–99 price rise. However, the likely cause of this "pause" in generally increasing demand growth rates was the fall in equity numbers on world stock exchanges, which triggered an erratic downturn in the world economy. To this can be added energy-demand-reducing impacts on world airlines, travel movements and consumer confidence in the OECD countries of the September 11, 2001, terrorist attacks.

Current oil demand growth rates in the Asia-Pacific region, second only to North America as an oil importer and consumer since 1992, are generally in the 5.5 to 6.5 per cent per annum range for most regional countries, including China and India,[9] and have tended to *increase* since 1998/99. Oil and gas import demand in this region is set to grow very rapidly, due also to localized depletion of current production capacity.[10] It is therefore easy to suggest that the "ten-year trend" annual growth figures proposed by the US EIA and OECD IEA, of 1.4 per cent for commercial energy and about 1.3 per cent for oil demand, must be an aberration. In addition, if oil prices played any role at all in setting this low growth trend, it was through *cheap* oil and gas in the 1986–99 period, which tended to reduce solvent international demand through reducing commodity prices and slowing economic growth rates of lower-income countries. This, in turn, reduced annual demand growth rates for commercial energy, and particularly for oil.

Generally, lower economic growth rates also applied, even in spectacular fashion, to the OECD countries in the 1985–2000 period. For the G-7 group of leading economies in the OECD bloc, average annual real growth rates fell by about 50 per cent, comparing average growth rates in 1989–95 with those for 1968–79, for numerous reasons.[11] This fall in average growth rates inside the OECD also resulted in slowed economic growth and falling oil demand growth rates for the "traditional" NICs and Asian Tiger economies (see Table IV.18.1, above), generally reducing world oil demand growth rates. Since at latest 1994–96, this overall trend (of about 1.3 per cent annual oil demand growth) has been replaced by a much higher trend, notably due to the "emerging" NICs with huge populations and immense markets, comprising not only China and India, but also Pakistan, Brazil, and Iran. These emerging economies have energy-intense economic activity. In addition, the aging and sluggish economies of the OECD are now experiencing major energy-economic change, including the replacement of sometimes very aged energy, economic and social infrastructures, markedly increasing the energy- and oil-intensity of their economic output.

One key example of this concerns the world's largest single oil consumer, the US, where oil demand through the first five months of 2003 increased by about 0.6 million barrels/day, representing a 2.9 per cent growth over December 2002, and a year-on-year growth rate of 2 per cent.[12] Combined with very firm demand growth trends in Asia-Pacific, it is most likely that low growth trends for both oil and energy have given way to higher annual growth rates. This is for a large number of reasons, which include energy infrastructure changes in the OECD bloc and the macroeconomic impacts of the "emerging" NICs, through their fast industrial and economic growth exerting a "pull effect" on the sluggish OECD bloc. We should also include the many and significant social, secular and cultural changes occurring within the OECD economies, which though almost unstudied from an energy point of view, almost certainly lead to a composite *increase* of their energy and oil demand.

The fact of oil demand shock operating from, at the latest, 1995 can be understood from the simplest and most aggregated figures, such as those shown in Table IV.18.3.

Table IV.18.3   World oil demand change by volume, % change on preceding year

1995	1996	1997	1998	1999
1.64%	2.15%	2.61%	0.52%	2.86%
1990	1991	1992	1993	1994
1.31%	−0.19%	0.51%	−0.04%	2.09%

Source: *BP Amoco Statistical Review of World Energy*, various editions.

PRICE SHOCK

The first "shock" is that there is an almost complete lack of price elasticity on a world-economic scale in response to oil prices that, through 1998–99, increased about 230 per cent. The argument made by this author of *reverse elasticity*, or an increase in demand when prices rise, is rather well supported by even these very simple aggregates. Taking the 1990–99 period, we can also note that almost every time oil prices tended to rise, *demand increased* within about six to twelve months (see Table IV.18.4). This is particularly clear for 1999 compared with 1998: after an approximate tripling in terms of peak trough yearly prices, world oil demand *increased* by 2.86 per cent over 1998, its highest rate in nearly a decade! Whenever prices fell during the 1990–99 period, demand growth rates tended to fall. This again proves, if further proof is needed, that world oil demand is dependent on global economy growth and yearly changes in that growth, and on many energy infrastructural, technological, energy-economic, social and cultural factors. Annual world oil demand is therefore usually *unrelated* to the oil price except when very,

Table IV.18.4   World oil demand and oil price variations 1990–99

Year	1990	1991	1992	1993	1994
Year min oil price 2003 $/bbl	20.75 US$/bbl	21.60 US$/bbl	21.50 US$/bbl	17.05 US$/bbl	16.90 US$/bbl
Year max oil price 2003 $/bbl	39.40 US$/bbl	34.55 US$/bbl	29.60 US$/bbl	26.65 US$/bbl	24.65 US$/bbl
Demand change % on year before	+1.31%	−0.19%	+0.51%	−0.04%	+2.09%
Year	1995	1996	1997	1998	1999
Year min oil price 2003 $/bbl	19.55 US$/bbl	21.05 US$/bbl	20.55 US$/bbl	10.95 US$/bbl	27.70 US$/bbl
Year max oil price 2003 $/bbl	25.20 US$/bbl	29.55 US$/bbl	28.15 US$/bbl	18.75 US$/bbl	28.95 US$/bbl
Demand change % on year before	+1.64%	+2.15%	+2.61%	+0.52%	+2.86%

Extracted from Table IV.18.5.

very high prices are attained in a very short period of time. Over the short term, and depending on prices attained, demand will often *increase* as prices rise.

WHY OIL PRICES CAN ONLY INCREASE

For a number of reasons oil prices have followed an erratic but upward trend since their most recent low, in 1998–99, of around US$10/barrel. The most recent "price shock" sequence can be described from various perspectives, including the following:

> It is useful to distinguish short-term price fluctuations from episodic movements that sometimes characterize certain longer periods of time. The most dramatic episode occurred fairly recently and is still very alive in people's minds: this is the 1998/early 1999 price collapse followed by rises which took prices to high levels throughout 2000. The WTI [West Texas Intermediate light sweet crude] price was at $17.65 per barrel at the beginning of January 1998. It reached a low of $10.80 in late December 1998, but the lowest levels were not hit until early February 1999 when WTI bottomed at $10.26 and Brent at $9.70. After that date the price movement was relentlessly upward with the WTI price ending the year at around $26.50 per barrel and peaking at $34.15 on 7 March 2000. It took 13 months of toil for the market to bring the price down by slightly less than $7.0 [a 39 per cent decrease] and then another 13 months of over excitement to raise it by almost $24.0 [a 233 per cent increase].[13]

Amusingly enough, Mabro and other commentators who characterize price increases as "over-excitement" and price falls as "toil for the market," trace the signal for this upward price movement to a late-1997 decision by OPEC to *raise* output quotas by 10 per cent. This in turn isolates a key element of oil market mythology – the fixed belief that OPEC has bottomless spare capacity, and will always have spare capacity, forever. For OPEC as currently constituted (including Iraq), and for the next three to five years, no reasonable analyst can go above 31Mbd to 32Mbd of exportable capacity, over and above domestic consumption needs. Speculation on this export capacity number is of course a prime subject of "OPEC watching," but many unbiased observers suggest the real maximum export capacity of OPEC today, and for the next three to five years, will have real difficulty exceeding 28Mbd to 30Mbd.[14] More importantly, overall exportable surpluses of current OPEC producers can only stagnate or diminish by virtue of geological necessity. The "key exceptions" of course include Saudi Arabia and Iraq (and perhaps Abu Dhabi, Kuwait and possibly Nigeria) in the OPEC group, and essentially the Russian Federation alone in the non-OPEC group of oil producers with large exportable surpluses that could be increased.

Oil market price setting, as Mabro and other commentators point out is a result of trading expectations, not facts. These expectations – in other words market mythology – conceal an underlying belief that there can only be slow, gradual and predictable rises in world oil demand, at the "old paradigm" rate of about 1.3 per cent per year. In addition, market mythology believes that supply from OPEC and non-OPEC players increases faster than oil demand. By consequence, prices spike from time to time, when demand very temporarily outstrips supply, but always return to very opaquely defined "normal" trading levels. For about 13 years, through 1986–99, these were set at around US$18 per barrel. Exactly how this price was first arrived at and then fixed is at least as opaque and mysterious as oil prices attaining US$100 per barrel, in 2003 money, during the Iranian Revolution of 1979–80, but perhaps relates to very cheap natural gas prices, operating a downward ratchet effect on oil prices. The theory behind cheap oil prices embodied in the lucubrations of M.A. Adelman – that the "right price" for oil is US$2.50 per barrel in 1972 money – has like Gresham's Law, fully displaced any consideration – theoretical or otherwise – of why prices should rise.[15] For a few weeks, from late 1998 to early 1999, the "right price" of Adelman was achieved, when prices in current dollars hovered around US$10 per barrel.

CHEAP OIL AND THE DEPLETION ISSUE

Any reasonably unbiased reader of the summer 2003 "depletion series" by the US *Oil & Gas Journal* would quickly conclude that oil and gas depletion, as ever, is a 40-year threat, and therefore a subject for the Keynesian long-term. Extremely large remaining and recoverable oil resources are claimed to exist in so-far underexplored or even ignored regions like the deep offshore South Atlantic region, in parts of Russia that would have been somehow overlooked, and of course in Iraq, where "real reserves" are claimed by some, mostly American writers to be far above 200Gb. World total endowment would, according to these optimists, be at least 4,000Gb, of which production to date is about 900Gb.

Much less is said about the "producibility" of these enormous but imaginary reserves – that is, the rate at which world annual oil production can be increased before some hypothetical maximum is attained, of perhaps 150 million barrels/day by about 2038 (a 2 per cent annual average growth rate for 34 years would bring world oil demand to 156 million barrels/day). Even less is said about oil prices. For the moment, most contributors to the *Oil & Gas Journal* depletion series appear to suggest, oil market traders will pursue the "toil" of talking down oil prices simply because supply outstrips demand, and cheap oil is so good for the economy. A host of expert opinion will always be on hand to opine this, lately using the approximate tripling of oil prices in 1998–99 as their explanation for the 2000–02 dotcom–telecom crash on world stock markets.[16]

The OECD IEA in its monthly oil market assessment, *Oil Market Report* for July 11, 2003, is constrained by the facts to record that world oil demand on an all-liquids base was running at an average of 78.08 million barrels/day in May/June 2003. Based on data in previous issues of *Oil Market Report*, this yields a yearly growth rate of at least 2.25 per cent for summer 2002 to summer 2003. Despite this, the IEA confidently forecasts that world oil demand will only grow by 1.28 per cent in 2003–04, attaining 79.08 million barrels/day as the rate of average daily demand by summer 2004. No explanation whatsoever is offered as to why world oil demand growth will now suddenly return to the "long-term trend" growth rate, after its "surprising" near-doubling. The IEA in its July 2003 report then goes on to offer the perspective of non-OPEC suppliers increasing their market offer by up to 1.7 million barrels/day in the next twelve months, leading to OPEC suppliers losing market share for a fifth successive year. The only explanation offered for the "Baghdad Bounce" in world oil prices is that OPEC has decided not to increase output, and that Iraq's oil output is only making a slow return towards prewar levels.[17] The now dramatic decline of North Sea oil production, with the UK and Norway losing a total of 0.516 million barrels/day capacity between June 2002 and June 2003,[18] and continuing gradual loss of US production capacity (a decline of 0.285 million barrels/day in the same period), while US oil demand has increased at a 24-year record rate of 0.6 million barrels/day in twelve months, are of course not mentioned by the IEA as factors raising prices.

The work of Deffeyes, Youngquist and the ASPO group[19] (see also Chapter 1) on real-world oil reserves and production potentials strongly suggests net additions to world production capacity will soon

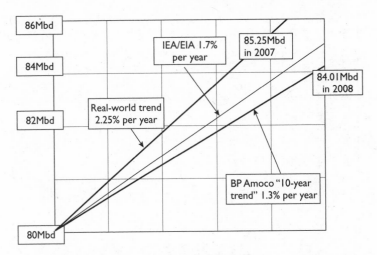

Figure IV.18.1   Real-world demand growth trend vs. IEA/EIA and BP Amoco.

fall to zero as the world arrives at its absolute peak of production. This will, through the deforming lens of the oil market, be tested in real time, and its effect will be vastly increased price volatility, followed by price explosion. After this, depending on the immediate economic reaction, some form of world compact to hold oil prices in a new and much higher price band will possibly be arrived at through hastily arranged "North/South" conferences like those of the 1974–81 period.

Some impression of possible new capacity required through the next five years can be obtained by comparing the three major trend rates of world oil demand growth discussed above. These are the current, real-world trend of about 2.25 per cent per annum (which may well be exceeded in 2003–04), the lower (1.7 per cent per annum) of the two trend rates used by the OECD IEA and US EIA, and the "ten-year trend" of BP Amoco (1.3 per cent), now resuscitated by the IEA in its forecasts for 2004 oil demand (growth of 1.28 per cent for July 2003 to July 2004). The variations, in terms of millions of barrels per day, with a potential for demand reaching about 87.2 million barrels/day in July 2008, soon become very large (see Figure IV.18.1).

CONCLUSION

For various reasons of economic doctrine cheap oil is seen by the decision-making elite in the richer nations as the passport to

Table IV.18.5 World annual oil price and volume changes, 1971–2003

Year average oil supply World: All liquids Thousand barrels/day	% volume change on yr before	Oil price Year min current dollars	Oil price Year max current dollars	Maximum oil price PPP adjusted 2003 dollars	Minimum oil price PPP adjusted 2003 dollars	Deflator	Deflator/California Energy Commission Delph IX Oil price forecast Survey State Govt, Sacramento, 1977 1996=100 1996–2003 Forecast average 2.9 per cent 3.4 per cent per year inflation
1971	50,785	4.45						
1972	52,540	3.45	US$1.84/bbl	US$3.70/bbl	US$15.50/bbl	US$7.75/bbl	29.36	
1973	58,505	11.35	US$2.35/bbl	US$3.75/bbl	US$14.80/bbl	US$9.30/bbl	31.21	PPP adjustment 2003/1973 = x 3.95
1974	58,610	0.18	US$11.60/bbl	US$15.50/bbl	US$56.15/bbl	US$42.05/bbl	33.94	
1975	55,690	−4.98	US$11.95/bbl	US$12.80/bbl	US$42.60/bbl	US$39.75/bbl	37.2	
1976	60,075	7.87	US$11.85/bbl	US$12.15/bbl	US$38.05/bbl	US$37.10/bbl	39.53	
1977	63,000	4.87	US$11.50/bbl	US$12.90/bbl	US$37.75/bbl	US$33.65/bbl	42.23	
1978	63,125	0.19	US$12.20/bbl	US$14.25/bbl	US$38.75/bbl	US$33.15/bbl	45.56	
1979	65,975	4.51	US$21.50/bbl	US$41.50/bbl	US$103.50/bbl	US$53.60/bbl	49.54	
1980	63,135	−4.22	US$28.50/bbl	US$34.95/bbl	US$79.65/bbl	US$64.95/bbl	54.2	
1981	59,745	−5.37	US$31.10/bbl	US$41.25/bbl	US$84.95/bbl	US$64.10/bbl	59.6	PPP adjustment 2003/1981 = x 2.06
1982	58,005	−2.92	US$28.50/bbl	US$34.75/bbl	US$67.80/bbl	US$55.60/bbl	63.31	
1983	58,040	0.06	US$27.45/bbl	US$31.50/bbl	US$59.05/bbl	US$51.45/bbl	65.88	
1984	58,650	1.05	US$28.60/bbl	US$36.50/bbl	US$65.25/bbl	US$51.15/bbl	68.82	
1985	58,150	−0.86	US$26.15/bbl	US$30.05/bbl	US$52.05/bbl	US$45.30/bbl	71.32	
1986	60,655	4.31	US$11.40/bbl	US$19.35/bbl	US$32.65/bbl	US$19.25/bbl	73.25	
1987	61,305	1.07	US$12.50/bbl	US$17.75/bbl	US$29.00/bbl	US$20.35/bbl	75.58	
1988	63,690	3.89	US$10.45/bbl	US$14.30/bbl	US$22.50/bbl	US$16.45/bbl	78.49	
1989	65,875	3.43	US$14.65/bbl	US$21.20/bbl	US$31.90/bbl	US$22.05/bbl	82.03	

Year							
1990	66,745	1.31	US$14.40/bbl	US$27.30/bbl	US$39.40/bbl	US$20.75/bbl	85.59
1991	66,615	-0.19	US$15.65/bbl	US$25.05/bbl	US$34.55/bbl	US$21.60/bbl	88.89 PPP adjustment 2003/1991 = × 1.38
1992	66,950	0.51	US$15.90/bbl	US$21.90/bbl	US$29.60/bbl	US$21.50/bbl	91.38
1993	66,700	-0.04	US$12.95/bbl	US$20.25/bbl	US$26.65/bbl	US$17.05/bbl	93.37
1994	68,100	2.09	US$13.05/bbl	US$19.05/bbl	US$24.65/bbl	US$16.90/bbl	95.34
1995	69,215	1.64	US$15.45/bbl	US$19.90/bbl	US$25.20/bbl	US$19.55/bbl	97.6
1996	70,705	2.15	US$17.05/bbl	US$23.95/bbl	US$29.55/bbl	US$21.05/bbl	100
1997	72,550	2.61	US$17.15/bbl	US$23.50/bbl	US$28.15/bbl	US$20.55/bbl	103.2
1998	72,920	0.51	US$9.70/bbl	US$16.60/bbl	US$18.75/bbl	US$10.95/bbl	107.9
1999	75,005	2.86	US$24.90/bbl	US$26.05/bbl	US$28.95/bbl	US$27.70/bbl	111
2000	76,905	2.53	US$22.95/bbl	US$34.25/bbl	US$36.40/bbl	US$24.40/bbl	114.3
2001	75,990	-1.19	US$17.60/bbl	US$29.55/bbl	US$31.05/bbl	US$18.50/bbl	117.6
2002	76,100	0.14	US$15.75/bbl	US$29.70/bbl	US$30.70/bbl	US$16.25/bbl	120
2003	78,700	2.75	US$25.00/bbl	US$37.5/bbl	US$37.5/bbl	US$25.00/bbl	123.5 PPP adjustment 2003/2002 = × 1.029
2004	81,500	3.44	US$35.00/bbl	US$55.00/bbl	US$55.50/bbl	US$35.50/bbl	127.5

OIL PRICES: *Oil Economists Handbook* (Vols. 1 and 2), G. Jenkins, Elsevier Applied Science, various editions; OPEC Bulletin; Platts Oilgram Price Report Prices are for selected volume crudes, including Saudi light, Nigerian light, Norwegian, US WTI, Kuwaiti light and other crudes.
YEAR AVERAGE OIL SUPPLY: BP Amoco *Statistical Review*; US EIA; *World Energy Statistics and Balances*, OECD-IEA, various editions; 1972–73 data is surely in error (11.3 per cent apparent increase, one year). Compiled by A. McKillop.

economic growth. This is a pure fantasy. Only at very high oil prices (between US$75 and US$100 per barrel) will the inflationary and recessionary effects of high energy prices be so strong as to cancel the global economic expansionary impacts of higher revenues for exporters of energy minerals and other energy-intense "real resources."

Since about 1995, "demand shock" has begun to operate in the world economy for a number of economic, social and technical reasons, leading to considerably higher underlying growth rates of world oil demand. One counter-intuitive or "perverse" reason for this shock is reverse price elasticity, or increasing oil demand with increasing oil prices. Current demand growth rates in world demand for energy and oil are about 2.25 per cent for oil and about 2.5 per cent for energy, on an annual basis.

Conventional economic growth will be enabled at the world level by oil prices rising to high levels, probably above US$60/barrel in today's money. This will serve to underpin, or even increase world demand for fossil-energy supplies, indicating that concerted international action is needed to plan for the accelerated arrival of Peak Oil, with Peak Gas being possible within ten to twelve years after that.

Because of depletion, but also because of environmental and climate limits, energy transition away from fossil fuels will inevitably occur. In the existing economic framework, price signals are necessary if this is to start, and to build sufficient momentum to be effective. Existing and developing frameworks provided by the Kyoto Treaty offer some potential for adaptation towards the task of energy transition.

Part V
After Oil

A long time before we reach "after oil," we will have oil (and gas) wars galore, and these are in fact going on right now – in Iraq, Afghanistan, Colombia, Chechnya, Sudan, Algeria and elsewhere. The big one to come, of course, is Saudi Arabia and possibly Russia; in both cases, at least initially, civil wars that will rapidly draw in other "players," simply because the consumer world cannot do without the oil and gas of these key exporters. Chapter 19 addresses the real point of these do-or-die experiments in "energy policy" of the muscular sort: these last oil wars will terminate with resource wipeout *and/or nuclear war*. The first is certain; the second, an option.

Probably the most dangerous myth that has been created in the last ten years is that of a "US hyperpower," a single nation with such unparalleled military strength that all potential or real foes will simply desist, and make treaties before the awesome might of its GIs armed with state-of-the-art weaponry, and of its unchallenged air superiority. An energy resource control war on the ground – probably in Central Asia – could line up the US, India, Russia, China and Pakistan, on one side or another, each a nuclear power. In all cases except Russia, all have conventional armies much bigger than that of the US. A war in Central Asia between these powers, for dominance over oil and gas resources, could not "go nuclear" until at least one of the players was beaten and driven out – because nuclear weapons' use in the oil and gas fields would sterilize the resources and infrastructures for producing and transporting the prize. So the conflict would have to be "conventional." This being the case, the US would almost certainly be beaten in a few months. Supporting evidence for this argument is found on the ground, every day, in Afghanistan and Iraq. The so-called hyperpower is unable to smash increasingly effective and deadly opposition within nations with not even one-tenth of its population. China and India are both vastly bigger countries than the US, and if necessary would use "human wave" tactics to destroy US troop concentrations completely, forcing the US to fall back on its so-called "air supremacy." Chinese, Indian, and even Pakistani missile technology is more than capable of decimating the US Air Force. Thus the US, whether it wanted to fight "conventional" or "high-tech" war, would be beaten and driven out. Its only solace would then be "resource denial," that is nuclear weapons use on oilfield and gas production areas taken by the winner or winners.

The administration of George W. Bush has acted to make such scenarios far from impossible. US troops are stationed in a swath of Central Asian countries and in vast numbers (around 200,000) in the Persian/Arab Gulf region. This in theory indicates "long-term presence and war readiness." We can only hope it is some popcorn-and-chewing-gum version of what an American military strategist might imagine to be Machiavellian intrigue and sophistication. If not, the countdown to perhaps terminal oil war could be figured in months rather than years. In any case, conventional war is an enterprise of extreme energy intensity. Western media gave breathtaking coverage to the marvels of US heavy bomber superiority in the shape of the few B-2 bombers the US has (real and serious long-distance bombing is accomplished by B-52s that are 30 or even 40 years old). The few B-2s able to be put on show made round-trip bombing raids to Baghdad from the US during the war of 2003, and used about 8 million barrels of aviation kerosene to make a few hundred sorties, only garnering a few dozen civilian fatalities in the form of "collateral damage," with few, or perhaps no, Iraqi military personnel being killed, while taking out "key installations" in downtown Baghdad. Since April 2003 the arrival of around 170,000 uninvited armed "guests" in Iraq has increased domestic oil demand by about 350,000 barrels/day. This has more than compensated for the near-total destruction of the already weakened Iraqi economy, which in late 2003 had an unemployment rate estimated by the UN at 65 per cent. The high-energy "migrants" into Iraq, according to the OECD IEA, had restored total domestic oil demand to *more than* the prewar level of about 0.6 million barrels/day by early November 2003. This is easy to understand – a US army Abrams tank, for example, needs about 3 gallons of kerosene *per mile*, and the US army's "Humvee" jeeps, much favored by Arnold Schwarzenegger in California (who claims to be someday converting "one or two" of his personal fleet to hydrogen), consume about 1 gallon for each 5 miles they patrol in occupied Iraq.

Sheila Newman takes on a courageous flight of imagination in supposing the world community arrives alive into the post-2035 world. Her article compares and contrasts France and Australia, after oil. In both cases the sustainable population is far lower than today's, dramatically so for Australia. In her analysis France might be able to pull through with no more than a 33 to 50 per cent reduction in current population numbers, about 60 million in 2003. The barren, thin-soiled continent of Australia, never having been glaciated and therefore never having accumulated significant soil thicknesses, will probably require 75 per cent reduction in population in order to be sustained without oil and gas (the last of which Australia has major reserves, but with little or no oil). How it does so, with successive governments eager to increase Australia's demographic weight and significance, and to play a major role in "world peacekeeping" as a supplier of troops to American oil adventures in the Middle East (and elsewhere), is at least open to question. As with the EU countries, in the US and Japan there will be an inevitable first stage of coming down to earth, and the collapse of the energy economy will precipitate years of grinding economic crisis. Not only our current political leaderships, but the consumer citizenry they represent will need hard-edged and incontrovertible proof there is no "magic bullet" or quick fix, and no way out of the daunting need to restructure both the economy and society.

After oil, we will certainly see the return of King Coal. We can note that the first ever international trade in coal, in Europe, dates from about 1695, and that coal was the fossil energy source for the Industrial Revolution. The article by Gregson Vaux is part of an ongoing research project of tracing a Hubbert-type curve for world coal. That is, the prediction of when we arrive at peak production and use of coal, which he tentatively places around 2025–40, with at least a doubling of the current world coal burn of just over 2.5 billion tons per year. In oil-equivalent terms, a total coal burn of 5 or 6 billion tons per year is not tiny, but is very small relative to what we obtain from oil and gas today, at around 35 per cent of current commercial energy from those sources. His basic premise is that "coal is there – it will be extracted and burned," and few can contradict this argument, despite the extreme environmental impacts that a massive increase in coalmining and burning will produce. In addition we can note that coal, at least as much as oil or gas, is a storehouse of chemicals readily available for the production of pharmaceuticals. While oil and gas will certainly be extracted to the last drop and cubic foot possible to maintain the consumer economy, the attitudes, policies and values of world and regional civilizations may have changed by 2035 or 2045. Retaining coal as a valuable raw material able to provide a swath of key pharmaceuticals for many centuries – if it is not burned – may become a policy that is not simply discarded the moment it is proposed. This hope cannot be completely eliminated from consideration.

As Sheila Newman suggests, a lower-energy, resource-conserving agriculture will by necessity be set in place in an After Oil Australia, as well as an After Oil France, using coal-based fertilizers in a much lower-intensity, more pinpointed approach to food production, processing and distribution. Australia's context of small, high-density city centers with vast, sprawling suburbs rooted in private car transport is the actual and emerging context in many countries. In spatial terms, suburbia takes a huge slice of many metropolitan areas, and Ted Trainer argues for restructuring suburbia. As his lively chapter clearly shows, a transition to sustainability in the sprawling suburbs of our existing cities is very possible. The major, most basic change that is needed, as ever, is cultural: of the values, goals and perceived needs of the individual. His "Simpler Way" is in fact similar to many people's lifestyles in special conditions – for example the suburban dwellers of big European cities during World War II. Another example is the forced intercommunal and collectivist, hyper-economical lifestyles forced on many suburban dwellers of Eastern European countries being adapted – through economic "shock therapy" – to the market economy. In many of these countries today, 25 per cent or more of the population lives outside the cash economy. In some ways these people are the first candidates for what Trainer argues should be an open and declared transition, given support and approval, their efforts facilitated by legislation, and not producing grinding poverty and marginalization. In fact, the victimization of such people has an almost declared goal – they have failed as candidates for the high-throughput economy. The fragile but repressive "culture" of that civilization cannot tolerate dissent or difference; its model of success is the only reality allowed today.

Both directly and indirectly, this book poses the question of *cultural values*. Western civilization is the product of demographic and environmental changes in the man–environment relation of the east Mediterranean, west Asian and north African regions through about 3000BC–AD500. The Industrial Revolution and the founding of the US, Canada, South Africa and other outgrowths of European expansion and colonization are all such recent events in historical terms that no particularly unique culture has formed. As Chapter 23 notes, the Roman Empire of about 500BC–AD450 at no time had anything we could identify as specific and unique culture. Like many imperia, all that the Romans had was borrowed directly from Ancient Greece and elsewhere in the region. In this sense, the Roman Empire resembles any of our "modern" civilizations, having only a flimsy mishmash of relic ideas and values as its cultural frames of reference, while its real mission was to conquer and expand, until it collapsed. The nearest thing to a real cultural foundation of Western civilization is therefore the Hellenic civilization. A few aspects of that civilization deserve special attention when discussing the collapse of the Western world's industrial urban civilization: in particular, the Ancient Greek cultural message regarding the relation between man and environment, where multiple layers of meaning are attached to various symbols and values through time, and the growth of human power over the earth and nature. The messages of Apollo and the Muses are even more relevant to us today: "know thyself" and "Moderation in all things."

19
The Last Oil Wars

Andrew McKillop

Without oil and the enormous fossil-energy-based infrastructure supporting it, and the social and cultural values enabled or forced by fossil-energy civilization, war will be very different from what we have seen since about 1900. Some war historians argue that the two world wars of the twentieth century may only have truly ended in 1991, with the collapse of the Soviet Union, or "Evil Empire," as it was called by former US President, Ronald Reagan. In other words, the twentieth century was a period of *permanent war*. The theory of permanent war, which can be traced back to late nineteenth-century historians, as well as Marx and Engels, has it that under certain conditions this "low-level war" occasionally breaks out into paroxysms of total war. For the twentieth-century version of the theory we have the numerous inter-imperial "brushfire" wars, preceded or followed by civil wars, during 1918–39. This then broke out into the paroxysm of World War II, and then subsided into the seemingly permanent Cold War of 1948–91. At the same time, through 1940–75 there was a nearly continuous anti-imperial "decolonizing" or North/South war. Thus, 1914–91 was indeed a period of permanent war, and already included conflict between larger and smaller powers for dominance over oil reserves. One key date in that period was that of the collapse of the Ottoman Empire in 1917. From 1917 to the next outbreak of total war in 1939–45 there had been major decline of the British and French empires, expansion of the Soviet Union, and growth of US economic and political colonialism, euphemistically called "influence." During this time, Germany lost its nineteenth-century colonies, but regained sweeping but short-lived global reach in the Third Reich of 1936–45. Mussolini's New Rome rapidly expanded through 1922–36, then collapsed with Nazi Germany. Japan's Co-Prosperity Sphere extended over much of East and Southeast Asia in 1933–45, and was then extinguished in the twin suns of Hiroshima and Nagasaki; while by 1949, China's future territorial and political ambitions had been set by Mao Tse Tung after a civil war costing tens of millions of lives. Marking the end of that 1914–91 phase in a near-century of war, some historians will someday see the so-called "Liberation of Kuwait," in 1991, as the turning point in a near-century of initially unabated warfare. But others may see this First Oil War as a model for, if not the *cause of*, those to come.

POPULATION, ENERGY AND WEAPONS

It is important to note that the expansion of population and energy supplies through the twentieth century was a one-shot event that will not and cannot be repeated – but the resulting "policy" or activity of total war will most certainly spill over to the first two or three decades of the twenty-first century: not only through rivalry for decreasing oil and gas reserves, but through the "stock" effects of human numbers and weapons supplies. From 1900 to 1999, world population increased from about 1.45 billion to nearly 6 billion, while fossil fuel production and consumption rose from about 1,100 million tons of oil equivalent (Mtoe) to some 9,800 Mtoe. The remarkable increase in the efficiency of use, with thermal energy actually rising from around 10 to 15 per cent in 1900 to over 25 per cent at the end of the century, resulted in actual energy effectively increasing close to 18-fold, corresponding to a four-fold increase in human numbers. Key indicators for output of both civil and military equipment such as automobiles and light artillery showed spectacular increases. While car production increased by about 45 times through 1900–40, the production of mortars, mines, grenades, service rifles and small caliber artillery (105-millimeter and below) increased about 200-fold in the same period. Nuclear weapons have spread, since the 1970s, from the Security Council club of five, to at least nine proud owner nations, but the existence of any nuclear reactor, anywhere, places nuclear-equivalent targets within easy reach for any enemy.

At the same time, world food production increased regularly and by large amounts until 1991, easily outstripping the rise in human numbers until the last decade of the century.[1] Peak annual growth of world population also occurred in the 1990s, at about 95 million in 1995, before declining to about 85 or 90 million in 2001–02, exhibiting all the characteristics of a new, long-term trend for declining annual increases. Population growth, right through the twentieth century, always rapidly compensated for any war losses. Estimates of perhaps 60 to 80 million war deaths in the 1914–18 and

1939–45 periods represent less than one year's *increase* in human population at the peak rate of 1995. All other war deaths and war-related deaths in the twentieth century – perhaps 45 to 75 million, depending on whether the deaths from epidemics and famine caused by war and conflict are included – represent about eight months of world population growth at the peak rate of 1995.

With abundant population, food supplies and weapons production were therefore in place for a century of war that will without any doubt spill over into the first few decades of the twenty-first century. Underpinning these factors was the discovery and production of oil and gas resources. US oil discoveries peaked in the 1930s, while discoveries on a worldwide volume basis peaked in the 1955–63 period, due to the intensive mapping, exploration, proving and production directed towards Middle Eastern oil and gas reserves that had already, in the 1917–39 period, been a focus for rivalry, competition and conflict within fossil-energy civilization. The Middle East and Central Asia, from now to about 2025, will without the slightest doubt again be the arena of a higher-tech, better-armed and more demographically numerous replay of what was called "The Great Game." In the earlier version, the main rivals were the European powers (notably France, the UK, and Germany), the US and the Russian (and then Soviet) Empire, as the Ottoman Empire collapsed in 1917, just before the defeat of Turkey and its German ally in 1918. Immediately on the retreat of Ottoman occupation forces, Kurdistan was proclaimed by its nationalist fighters, then recognized by and represented at various conferences and meetings held by the US, British and French victors of World War I, known as the Versailles Treaty series, and continuing through the period 1917–23. By 1922, and unquestionably because Kurdistan was known to have a very large proportion of all oil discovered and proven in the region at the time, stretching from southeastern Europe to the Indian Ocean, Kurdistan was simply "de-recognized" by the US, British and French victors. They then split it up between Turkey, Iraq, Syria and Iran, all of which had themselves been subject to considerable frontier modifications, or had never previously existed as national entities. Another and later national victim, in this region of influence-trading and map-drawing exercises in imperial tea rooms (always behind closed doors), was Palestine. Iran, too, was for several decades a bone tossed in the air for Soviet, British, American and national players to dispute, to draw "definitive" frontiers for, and in which to find and produce oil and gas.

GREAT GAME II

With the collapse of the Soviet empire in 1991, after its symbolic defeat in Afghanistan by US-backed mujahidin including Osama bin Laden, the accelerating decline of oil and gas discoveries, accompanied by rising import demand in East and Southeast Asia and the Indian subcontinent, brings new nuclear-armed players into what will become the Great Game II. Imperial map-drawing exercises no longer carve up strategic resources in the tea rooms, but in the air-conditioned think-tanks and bunkers of Washington, Paris, Beijing, London, Moscow, New Delhi, Islamabad, Tehran and elsewhere. Troop strengths in the broadly-defined Middle Eastern and contiguous regions, containing about 55 per cent of all remaining world oil reserves, have been multiplied at least six-fold since 2000. Great Game II is in fact a continuation and intensification of the previous round begun in 1917, which continued through World War II, became interlocked with the birth and expansion of Israel, triggering the Iranian Revolution of 1979, and then gave birth to the 1991 "liberation of Kuwait," or Oil War I. Great Game II focuses on a broad sweep of oil- and gas-bearing territory and contiguous pipeline routes stretching from ex-Soviet central Asia to Iraq, itself incorporating much of "de-recognized" Kurdistan, and continuing with the oil-bearing source rocks to Saudi Arabia, Kuwait and the other oil-producing states of the Gulf, facing Iran.

It may be a surprise to some policy analysts supplying analysis for Great Game II strategists that Iran, which was the "Persia" in the Anglo-Persian Oil Company of the 1920s and later became BP (which by the late 1990s had adopted the nickname "Beyond Petroleum"), is an oil exporter long past its peak production capacity (probably 1978), while its national domestic demand continues to expand with population and economic growth. Iran is so far past its peak that by 2008–11, according to its own ISIR scientific agency,[2] Iran will probably cease to have any exportable oil, and will become an importer country. For Iran, other than the siren call to develop nuclear power – opening the way for nuclear arms and providing a sure footing for standing firm against Israel – the real and effective solution will be to develop gas-to-oil conversion (GTO), enabling Iran to produce synthetic oil from its immense gas resources. Even before the overthrow of Shah Pahlavi by the Khomenei-led revolution of 1979, the country's oil discovery and production indicators revealed that Iran was already heading towards that day – then a long way in the future – when it would cease to export oil. One consequence was Iran's adoption of a program to develop nuclear power for electricity production. As has been amply proved by India and Pakistan, and most recently North Korea, *any* country with civil nuclear power is at most two screwdriver turns from nuclear weapons capability. Great Game II strategists, despite their air-conditioned bunkers, might toy with "Indiana Jones" images of those imperial tea rooms, of Sopwith biplanes and Mauser and Remington rifles left over from the 1914–18 war; going too far in their reverie, they might imagine they have the *time* that Great Game I players had, as they inched forward in their strategy, through occasional armed skirmishes against lightly armed and disorganized enemies, to find and keep abundant and cheap oil and gas resources in the Golden Triangle, centered on what was to become Saudi Arabia. The replay will be different, and will come to an accelerated end. It has only two variants: resource wipeout with, or without nuclear war, the first of which is inevitable, while the second remains an option.

A DIFFERENT WORLD – NEW DANGERS

Saudi Arabia and the Arabian Peninsula of the 1920s differ substantially from the Middle East and Central Asia of 2003; population numbers and weapons stocks are vastly greater today. But there is one geopolitical similarity with that long-gone period of languid, oil-seeking intrigue in the desert sands. The causes of World War I may be debated endlessly, but most historians agree that cracking geopolitical faultlines in the Balkans and southeastern Europe, together with the fall of Tsarist Russia and the emergence of the Soviet Empire, made total war inevitable. Today's shrunken and impoverished Russian Federation, despite its impressive nuclear missile stocks, has withdrawn over the horizon; this vacuum is now more than amply filled by the oil-hungry US, with a troop presence in each of the former Soviet Union's southern and Muslim republics, and with the military occupation of Afghanistan (and now Iraq) after their invasion, the overthrow of the indigenous regime, and the installation of some puppet figurehead – such as Hamid Karzai in Afghanistan, a former employee of Vice President Cheney – as head of state. Afghanistan shares a frontier not only with nuclear-armed, oil- and gas-importing Pakistan, but also with China, whose fast-growing dependence on imported oil will force Beijing's strategists to participate in Great Game II.

In Soviet times, tracing the frontiers of the USSR's southern republics was at best haphazard, and massive deportations and forced population movements were carried out to comply with Moscow's political whims and fantastical economic plans. A key example were the massive deportations from Chechnya, the martyred "pipeline-route state" of today. The lifetime of leftover regimes in the Former Soviet Union's Muslim republics, like that of the integrity of leftover frontiers from Soviet times, is probably short. The entire checkerboard of geopolitical power and influence rests on fragile, hastily drawn frontiers which, today, relate to almost nothing. Great Game II players must contend not only with these artificial frontiers, but also with the aftermath of a primal clash between Marxist values and the call of Islam, now that Cold War inertia has been so rapidly and totally stripped away. This, together with the accelerated timetable that Peak Oil sets for winning and then burning those lifeline supplies of precious oil, creates almost open-ended risks in a very dangerous but certain race to oil wipeout – with or without nuclear war.

20
Future Settings: Perspective for Sustainable Populations "After Oil" in France and Australia

Sheila Newman

This chapter explores and contrasts the potential for human survival in Australia and France after the petroleum interval, assuming that this interval may have lasted between around 1850 and 2035. To do this it is first necessary to establish pre-fossil-fuel carrying capacity, assuming that soil, water and climate retain their productiveness. The carrying capacity sought for France is as a self-sufficient agricultural country, with a high proportion of cereal crops. In conceptual terms, a return to later nineteenth-century land use and farming intensity, if not techniques, and to early to mid-nineteenth-century population numbers, may not be as difficult as it might look, because we need only to go back to the past. For Australia the method is different, because the people there prior to European settlement produced no recorded history. It is also difficult to compare pre-fossil-fuel European agriculture with that in Australia, because much of the land was developed after World War II. To work out Australia's productivity it is necessary to use paleontology, archeology, anthropology and ecology. Fortunately, Australian scientists have pioneered the collection and analysis of these kinds of data for just this purpose.[1]

Paleontologist Tim Flannery has popularized the notion of vastly greater carrying capacities in Europe than in Australia by referring to calculations of biomass in both regions.[2] For instance, he points out how Europe easily sustains a vast population of humans plus 27 species of mammalian carnivores, including two species of bears – which are the biggest, most energy-intensive mammal – in an area not much bigger than Australia.[3] Australia, of course, supports a much smaller, less differentiated biomass. The reason for this difference, apart from good rainfall, is Europe's soil resources and the role of glaciations – major glaciers only having completely withdrawn in the last 8,000 years, grinding and renewing the earth as they moved.

Such major geological events are responsible for Europe's rich and thick topsoils reaching depths of eight or nine feet.

FOSSIL-FUEL POPULATION SETTINGS FOR AUSTRALIA AND FRANCE TODAY

Presently, Australia's population is about 20 million, and since 1990 it has increased by 2,198,550 people, or approximately 12.8 per cent for an average growth rate of +1.6 per cent per year. Almost half of this increase was due to immigration. In 2003 Australia was on course to reach about 30 million by 2050. By contrast, France's population is 60 million and, since 1990, had grown by 2,462,700 people, for an average rate of +0.39 per cent per year. A large part of this growth was due to natural increase. After 2050, with the demise of most baby-boomers, France's population is set to decline.

Australia

In contrast to the rich, deep European soils of France, Australia's topsoil is often only a few inches deep, if that. The biophysical constraints of this, the oldest continent, are largely due to its flatness and lack of recent major geophysical upheaval. As well as affecting the potential for rainfall, the flatness means that rivers flow slowly, little silt is gathered or deposited, and salt accumulates in the soil. The absence of volcanoes, earthquakes or glaciers means that the soil fails to be renewed through widespread grinding and crushing of minerals and rocks. Successful non-nomadic Australian mammals tend to be small, with unusually slow metabolisms. The largest mammals are nomadic macropods – the kangaroos – which can travel quickly away from drought to rain, to find better foraging. Many of the largest fauna are land-based reptiles, requiring low levels of food-energy, which use heat from the sun to raise their body temperature.[4] Nomadic birds, like cockatoos, parrots, and the flightless emu, also do well. Both flora and fauna display numerous adaptations to precarious, stingy soils and an arid climate. Some of their many unusual features are the number and variety of adaptations to extreme soil infertility. These include those of the world's largest variety of carnivorous plants, which are thought to supplement nitrogen-poor soils with insects, and a tree-sized saprophytic mistletoe that draws nutrients from the roots of other vegetation.[5] The sclero-morphous structure of much indigenous vegetation is an adaptation to water shortage.

In Australia, pre-fossil fuel society was hunter-gatherer based, averaging on continent-wide terms less than one person per 8.5 square kilometers – possibly as few as one person per 51 square kilometers.[6] There was no agriculture, almost certainly due to the climate and soils.[7] To get an idea of what this means requires an understanding that the majority of the continent is hot desert.[8] Total land stock is 770 million hectares (7.7 million square kilometers) but under 30 million hectares, or 300,000 square kilometers (less than 4 per cent), is of good or very good quality, in terms of the range of its cropping potential.[9] Rangelands encompass some 75 per cent or 570 million hectares of the continent. About 406 million hectares are used for grazing, with stock density running as low as one beast per 100 square kilometers. Rainfall is highly variable, with frequent droughts lasting several seasons, resulting in massive die-offs.[10] Like the nomadic adaptations of kangaroos and birds to erratic climate, the Aborigines, who were hunter-gatherers, moved with their food sources in response to changing conditions, with the exception of those hunting and gathering under more settled conditions in comparatively more fertile, less arid parts, especially the southeast. The continent supported numerous clans at different densities, according to regional soils and climate. The distribution of the fossil-fuel-era population is similar, also reflecting climate and soils, although abundant fossil fuel has made it much denser and more numerous.[11]

Net primary productivity, continent-wide, was about 5 per cent less prior to European settlement in 1788.[12] For all intents and purposes, this increase has been restricted to about one-quarter of the country's surface area, mainly in the southeast. From this we can infer that net primary productivity has increased in this more fertile quarter of the continent by about 20 per cent, and that productivity in the most fertile 4 per cent would have increased by over 100 per cent. Almost all of this is due to the application of fossil-fuel-based synthetic fertilizers, irrigation pumps, and fuel-burning engines for machinery and transport.[13] Synthetic fertilizers, irrigation and mechanized agriculture are beginning to produce serious diseconomies, notably by triggering the desertification of previously productive areas.

Prior to European settlement, for at least 40,000 years, Australia was occupied by clans of Aborigines totaling a population estimated to have been between 150,000 and 300,000, although some estimates have put it as high as 900,000.[14] With wood, wind, draft animals and some coal the population rose to around 6 million prior to World War I. This is the number that Flannery thought would be sustainable in the long term for Australia, with a relatively comfortable lifestyle.[15] He was, however, probably also thinking of continued fossil fuel use. It is likely that long-term carrying capacity without substantial quantities of fossil fuel may be closer to that of the pre-European Aboriginal population.

The question of determining the carrying capacity[16] of Australia, as a pre-fossil-fuel, self-sufficient agricultural economy, is problematic on many levels. None of these, however, should stop discussion. Firstly, the notion of an agricultural economy is counterintuitive because of the inherent unsuitability for cultivation of most regions and soils of Australia, and because of severe, widespread damage to soil and water resources since European settlement. Secondly, there is the certainty of further and massive land degradation induced by ongoing processes. Dry-land salinity (currently irreversible) affects 2.5 million hectares; some 17 million hectares of the 30 million total of good land are likely, on current trends, to be destroyed by salinity by 2050, leaving 13 million "good," and leading to a probable halving of agricultural productivity.[17] More than 24 million hectares of soil is considered acidic. Much of this is natural, but agricultural management technologies are causing the soil acidification process to accelerate.[18] The distances and areas required for traditional agriculture are so vast that they are generally incomprehensible to those from the Northern hemisphere without science-based ecological knowledge.

In performing a ballpark calculation of carrying capacity[19] after fossil fuel, we need to consider the following. With productivity of land approximately halved by 2050, and a population of 30 million (about 50 per cent above today's), Australia's export economy, based on agricultural and mineral products, will shrink unless an elite can maintain exports through the intensified exploitation of the local population. We can assume effective depletion of both oil and natural gas by 2050.[20] Taking into account its role of substituting for declining oil and gas, world coal may last (depending on various factors) up to around the middle of this century, or, in the unlikely case of zero consumption growth, up to the middle of the twenty-second century.[21] Conventional nuclear energy sources, if utilized in Australia, will possibly last until around 2100.[22]

Returning to the notion that an agricultural economy is counter-intuitive, we can suggest that the optimal survival system will be a hunter-gatherer, herding economy. This will include some "oases" of gardening and crop production, offering the most natural, logical and ecologically efficient solution to a low-energy future. In Australia we have a natural biodiversity that has adapted beautifully to the biophysical restrictions of the Australian continent. It will make far more sense for humans to adapt to and operate with this biodiversity than to continue with uncontrolled and uncoordinated modification of the environment, flora and fauna, using the "big stick" of fossil-fueled intervention.

After the fossil-fuel interval, the continent's capacity to support more than its natural biomass will necessarily reduce to near zero, and the human population will shrink, one way or another. If we assume a loss of at least half of the 5 per cent gain in agricultural productivity since 1788, then we are perhaps contemplating a population 2.5 per cent larger than the aboriginal population before 1788 – that is, *below 1.5 million*. Such a population would subsist mainly by hunting indigenous fauna (like macropods and birds) or herding exotic, imported fauna (like cattle), using draft animals (camels, horses and cattle), and cultivating, by recycling manure and other wastes, a greatly reduced area of arable land, and using flow energies, of which the most dominant will probably be wind,[23] with some solar and some biomass if combustion engines are maintained for limited, specific purposes. These flow energies might add to the productivity of the land, but from that gain should be subtracted land needs for work animal fodder, and the human built environment.

It is unlikely that this tiny population would benefit from large-scale hydro-electricity systems, except perhaps on the island of Tasmania, which has some fast-flowing rivers. Inland water sources on the mainland, with the exception of the unreliable Murray–Darling River system, are almost all unable to support fluvial transport and reliable power generation. Wind and draft animals were the power sources most used prior to fossil fuel in Australia. Geothermal, solar, and tidal energy offer limited opportunities, but require high technology for sophisticated harnessing, and may not be practical for a small, post-fossil-fuel population living largely off the land. Failing massive technological breakthroughs, these sources are not likely to improve greatly on the continent's original fertility. Note that I have not talked about use of the existing transport infrastructure, as I will do for France. This is because, although it might be possible to use trains and grid electricity on a limited scale, the distance between cities and the fall in population makes maintaining these options unlikely. There might perhaps be a case for wood-fired rail transport from the major inland food production area (should any of this survive), with secondary distribution by road using draft animals, and coastal shipping.

The Australian planning and development system, and its construction industry, have successfully fought necessary policies to reduce population growth and energy demand. Australia's inability to plan and adapt infrastructures, industry and national resource needs to radically changed future conditions is a major problem. The federal government lacks authority to direct, oversee and coordinate state and local government land-use planning.

France

Although Europe and France were blessed with very rich soils and a climate conducive to agriculture, it must be noted that modern agricultural equipment, irrigation, single-cropping and a near-total reliance on mineral fertilizers has radically increased erosion and soil degradation[24] all over Europe, this being compounded in regions north of about 45°N latitude by increasing rainfall, probably due to global warming.

From medieval times until the middle of the eighteenth century, France's population oscillated around 18 or 20 million, which at the time was the largest in Europe. Growth was then relatively rapid, attaining nearly 30 million by 1815.[25] Territorial expansion through warfare also increased France's population by altering its borders: in 1850 nearly 1 million more people and their territory, in the form of Nice and Swiss Savoy, were added. Since the middle of the nineteenth century, however, increased agricultural land due to drainage, irrigation and other works was more than counterbalanced by loss of better land to urbanization, soil degradation, and pollution from both industry and intensive agriculture. Despite this, it seems possible to consider a population of around 20–25 million as sustainable in the post-fossil-fuel era. Further, the policies and practices applied regarding the conservation of soils, the recycling of soil nutrients, and the recovery of biological diversity may enable France to reconstitute soil quality and restore pre-nineteenth-century productivity in some regions. In this case, it might be possible to maintain the higher end of the very rough, sustainable population estimate given above.

France did not begin to experience its own industrial revolution until around 1880. World War I, the Great Depression, and World War II further delayed this development. There was little local coal, except in rather isolated and restricted areas. Probably for this reason, but for others as well, French population growth was modest relative to those of Germany, the UK and Italy. Between 1815 and 1845, France's population grew from 29.4 million to 35 million, largely due to skilled immigration. Subsequent growth still remained lower than for other large European nations, with the French national population only increasing from 35.6 million to 38.4 million through 1850–69.

In 1869, when the population of France had reached 38 million, horses, donkeys, oxen and even cows and dogs were used for road transport and hauling. The dominant industrial energy sources were still the water wheel, windmills and tide mills. Rivers provided the most energy-efficient form of transport and were extensively used before the fossil-energy period, with some transport networks connecting to those of other European countries. Bulk transport, wherever possible, was by boat and barge. Today, and excluding large-scale hydro-power (producing about 70TWh/year), small-scale (below 150kW) and run-of-river hydroelectric installations produce about 4.5TWh/year. France also has one of the world's few operational, large tidal electric power stations (Rance River, Brittany). In the late nineteenth century coal was increasingly used, but wood, which still provides about 40 per cent of space-heating fuel requirements in rural areas, was a major source of both commercial and non-commercial energy.

Much of the pre-fossil-energy infrastructure either exists or could be reconstructed, or even improved upon. This is particularly true of the canal system and woodlands. France's woodlands and forests have been maintained, even increased on earlier times. Efficiently used, in combined heat-and-power facilities, wood and other bio-mass energy resources can easily provide reasonable heating, and sensible electrical energy needs of an equilibrium or sustainable population of around 20–25 million people. The capacity of the managed forests of France to support the return of a functioning biodiversity, from which people could supplement food sources (small game) and obtain fuelwood, will however require big changes in the vegetation species mix for best adaptation to regional climates and soils. In the future, there may also be potential for wind-powered transport, both along canals and rivers, as well as roads.

In the late nineteenth and early twentieth centuries, the primarily rural population of France, employing candles and mineral- and animal-oil lamps for light, was often little integrated into the growing urban and industrial-based cash economy. The peasants had acquired land ownership rights through the French Revolution of 1789. This relatively secure peasantry, with strong cottage industries, lacked the motivation to provide the "factory fodder" of the dispossessed, as in the UK and other European countries where the people were still serfs or had lost all title and communal access to land by the Middle Ages. The French were similarly reluctant to settle France's colonies. Some of the nineteenth century's demographic growth may have been supported by wealth, or related commerce, arising from foreign possessions, particularly in the cities.[26] Without question, access to land, food, and energy-producing resources will form an important part of those "social contracts" that new regional or national entities will develop in the period from about 2035.

Primary productivity gains are dependent on the fossil-fuel economy. While the French may claim the status of "the EU's breadbasket" this ignores two key factors. The first is the pollution and depletion of water tables, most spectacularly for pollution in Brittany, and most intensely for depletion in the entire south and southwest of the country. Secondly, although French agriculture is among the most productive in the world, it is heavily dependent on fossil fuels and petroleum products – both directly for machines, and indirectly for fertilizers, insecticides, animal medication and other inputs vitally necessary for intensive production. Once these are stripped away, food self-sufficiency for perhaps 25 million people – less than 50 per cent of France's 60 million population in 2002 – becomes an optimistic but perhaps attainable target.

It should not be forgotten that France was not really a political, linguistic or cultural entity before the early nineteenth century. The development of railroads, post and telegraph systems played a major part in the sudden rush of "nation-building" that occurred in Europe through 1780–1860, and France provides an excellent example of this. The ability to cover wide areas with the fast transport, and now communications, necessary to bind disparate communities and cultures into what are called modern nations would necessarily diminish fast with the loss of fossil fuel. But it might be possible for France to maintain electricity supplies for a certain level of high- or medium-speed rail transport[27] through a mixture of renewable energy sources – that is, water, wind, wood and biomass, and tidal energy – although nuclear electricity will most certainly be the first choice of current policy makers.[28] The infrastructure is also still largely in place for canal traffic at certain levels of capacity (a few per cent of current road transport capacity). The spatial organization of human settlements, outside the unsustainably large and energy-intensive cities, includes certain amounts of building stock capable of being adapted to much lower energy operation, and therefore utilization.

The twentieth-century infrastructure overlay to the above target population (20–25 million), which is the same as that at the beginning of the nineteenth century, offers some potential for restructuring, adaptation and continued use. This notably includes rail transport. At present, some all-electric, but mostly diesel-fueled onboard electric generator-powered trains link major points of the country and neighboring countries. Other than canal and maritime transport, rail transport is the most energy-efficient. Building stock, insulation and design improvements could be applied to selected building groups in efficiently and rationally located settlement centers *outside* today's major urban areas, such as Paris – Ile-de-France, Marseilles, Lyon, Lille and Bordeaux, enabling a stabilized and decentralized population to live in an increasingly sustainable way. Most areas of France have inter-settlement distances set by pre-fossil-fuel "nested population hierarchies,"[29] which reflect distances easily traveled on horseback, on foot or by boat. It should be possible to return to the use of horses and other beasts for transport. If all else fails, it might even be possible to use animals, wind power and water power to draw light trains along rail lines, and for city transport to include biomass electric or horse-drawn trams such as were used in the nineteenth century. The critical question will remain the *maintenance of energy-intensive infrastructures*, which at present are entirely, or substantially, dependent on fossil fuels.

21

A Projection of Future Coal Demand
Given Diminishing Oil Supplies

Gregson Vaux

This chapter describes preliminary work that questions assumptions about the lifetime of coal reserves. Its purpose is to determine how long these reserves will last assuming that oil production will peak in 2009, and that its decline will follow a bell-shaped curve (Hubbert curve). It is further assumed that the energy currently provided by oil will be replaced by coal as oil production declines. Four scenarios were calculated, assuming world economic growth rates of 0 per cent, 1.5 per cent, 2.2 per cent, and 3.1 per cent respectively.

A number of groups and institutions, including the US Department of Energy, have devoted substantial effort to determining when the world production of oil will peak. Dates range from 2004 to 2112, while the USGS persists with ultimate oil reserve projections enabling it to claim a peak year of 2037 as being possible.[1] For this study, a world production peak is assumed to take place in the year 2009, although other possible dates for the peak would not substantially alter its conclusions.

After the world oil peak, there will be an inevitable decline in yearly production as individual wells throughout the world are depleted. Although the actual year of the peak is a matter of debate, it is certain that oil is a finite resource; when it becomes more difficult to produce, other sources of energy will need to be found. Although several alternative sources are proposed, such as solar and renewables, or nuclear, or coal, each has its problems and it is possible, indeed likely, that none of them *except for coal* will be an adequate substitute for oil.[2] Coal is an abundant resource, is widely used and has a well-established infrastructure, and is often described as existing in reserves that will last hundreds of years at current rates of use. Although world coal reserves are truly vast, it should not be assumed that demand for, and use of, coal will remain constant. If the world economy is to grow, then more energy will be needed, and when oil begins to decline, within the next few decades, it will cause an even greater demand for coal. Reserves now described as "worth hundreds

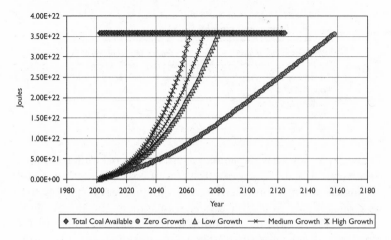

Figure V.21.1 Cumulative coal demand.

Source: Energy Information Administration – US Department of Energy.[5]

of years" could actually be good only for decades. But if it is assumed that declining oil supplies are to be replaced by coal, and if it is further assumed there will be economic growth, then coal demand under these conditions can be predicted, along with the time-frame in which coal reserves will be depleted.

The US Department of Energy has predicted that over the next two decades, world energy demand growth will increase annually from 1.5 per cent to 3.1 per cent in response to a growing economy.[3] Thus values of 0 per cent, 1.5 per cent, 2.2 per cent, and 3.1 per cent annual growth rates were used in order to calculate future energy demand. Further, all decline in oil availability is assumed to be compensated for by coal. The total world supply of coal in all its forms is assumed to be 1,080 billion short tons, this quantity having a total fuel energy value of 3.6×10^{22} joules.[4] In reality, if coal is to replace oil, then there will be losses from conversion of solid coal to liquid or gaseous fuels, but in this study these losses are ignored. Figure V.21.1 shows the yearly cumulative demand for coal after an oil peak in 2009. The horizontal line illustrates the total maximum supply of world coal that is currently assumed to exist.

The curves show four possible rates of growth and lifetimes for total coal reserve depletion. Table V.21.1 shows the year when coal will be depleted in the different scenarios.

Table V.21.1 Dates of world coal depletion given reduced oil supplies and economic growth[6]

Rate of economic growth	Year of total coal depletion
Zero (0%)	2158
Low (1.5%)	2080
Medium (2.2%)	2071
High (3.1%)	2062

The dates for coal depletion assume that coal can be mined relatively easily. In reality, the last ton of coal will not be as easy to mine as the first ton taken from the ground. It is very likely that as coal is mined from ever deeper seams, more and more energy will need to be expended to bring it to the surface. Where this is the case, coal prices will steadily increase, and this will result in falling supplies.

FURTHER STUDY

This chapter only gives a brief overview of how we can determine demand growth for coal, and how this will change the lifetime of world reserves. Future coal demand will be based not only on economic growth but also on declining oil supplies. Although much work has been conducted in the area of oil reserves, additional investigation is still needed. Further, I expect to expand the scope of this research to profile individual country coal production and supply situations. As energy becomes more scarce, it will be less than surprising to find that some or even many coal-producing countries will lose interest in exporting and selling their energy resources on the free market, and will decide to keep them for domestic consumption. With this in mind, supply and demand curves will need to be determined for individual countries, and the coal *price* function will need to be integrated into ongoing study. We will probably see that countries with the largest coal reserves will have an easier time adjusting to diminishing oil production. The countries with the seven largest coal reserves are listed in Table V.21.2.

A final assumption is that coal will remain as easy to mine in the future as it is today. In reality, more energy and money will need to be expended to mine coal in the future as the remaining deposits will be sparser, in less convenient locations, and deeper in the earth. This means that coal production will peak and decline *in much the*

Table V.21.2   Countries with greatest coal reserves

Country	Coal reserves in billions of short tons (approx.)
United States	273
Russia	173
China	126
India	93
Australia	90
Germany	73
South Africa	54

Source: Energy Information Administration – US Department of Energy.

same way as oil in the continental US and other oil provinces of the world, notably the North Sea, Indonesia, Iran and Venezuela.

THE SHAPE OF A COAL FUTURE

In the future, as our energy sources dwindle, we will rely less on oil and natural gas, and more on coal, and sources such as nuclear, wind, and other renewables. It is my belief that, for two or three decades after the peak in oil production, coal will be the world's dominant source of power. Coal will be burned in power plants to generate electricity; it will be transformed to gaseous fuels to replace natural gas; and it will be liquefied in order to be burned in internal combustion engines. In short, coal will be able to replace oil in many ways – but at a cost. First of all, the process of transforming coal into liquids and gases will be at an energy cost, which will mean that it will be more expensive than the fuels we obtain from oil and gas. Secondly, coal has a traditional and deserved reputation for being dirty.

I grew up in the steel-producing city of Pittsburgh, Pennsylvania. I remember, as a child, riding in the car with my parents past the steel mills. Coal and iron had made our city great, but had also made it filthy. On many days the smell of sulfur would hang heavily in the air and the sun would be blotted out by the black and yellow clouds that hung over the buildings. I don't remember it myself, but I am told that the workers in the office buildings would bring an extra shirt each day so that they could change into a clean one at noon, because the coal dust would make everything dirty so quickly. The remnants of the coal dust can be seen even today. Any building in Pittsburgh that is more than a few decades old is either black or has been cleaned using pressure washers or sandblasting. The church that I attend had been black for as long as I could remember, until it was cleaned two years ago, to reveal that it was actually built from cream-colored sandstone.

Burning coal has the potential to create not only black clouds and hellish odors, it also creates sulfur, nitrogen, and mercury compounds, and releases significant radiation that can have serious health consequences. The good news, however, is that most of these pollutants can be removed using modern technology – but at a serious cost. The scrubbers, activated carbon injectors, and baghouses (filters) that clean the combustion fumes are effective, but expensive both in lost energy and materials consumed. In the US there is currently a push to remove mercury from the exhaust gases along with additional sulfur and nitrogen compounds – but the energy producers complain that the cost of electricity will go up by as much as 5 per cent, or US$2 billion per year.

It is very likely that more coal will be burned in the future, much more. However, we cannot know whether it will be done in a clean manner; the necessary technology does exist, but it is based on our current economy. It is not hard to imagine that after oil production has peaked, demand for energy will be so great that pollution concerns will be seen as a luxury of the soon-to-be-past Golden Age of Cheap Energy. Who can afford to clean coal exhaust when children are going hungry or their families are shivering in unheated lodgings? In the days before pollution became a concern, there was a phrase that could be heard in the Pennsylvania coal towns. When visitors would complain of the sulfurous smells coming from the mills, the town residents would say: "Do you know what that smell is? It's the smell of wealth." We may well see the day when black skies and choking fumes will be a much-preferred fate compared to the poverty that comes from permanent energy shortage.

22
The Simpler Way

Ted Trainer

Before discussing energy arrangements compatible with an ecologically sustainable society, it is necessary to clarify the extent to which rich countries presently exceed the levels of consumption that could be maintained indefinitely, or extended to all people. The overshoot is enormous and, accordingly, the amount of energy use in a sustainable society will have to be a small fraction of the amount we take for granted in consumer society today. It follows that a sustainable society cannot be achieved without very radical changes in lifestyles, systems of land use, patterns of settlement, the economy, and social values.

The following is only the briefest review of major arguments supporting the "limits to growth" outlook on our global situation:[1]

- If energy production was increased to the point where a world population of 9 billion people consumed energy at current per capita rates of the rich world, *all* estimated fossil fuel reserves (*including* an assumed 2,000 billion tons of coal) would be totally exhausted within about 40 years.
- The "footprint of productive land" required by each person to sustain a rich-world lifestyle is around 7 to 12 hectares. The per capita amount of productive land on the planet is only about 1.2 hectares, and by 2050 will probably be close to 0.8 hectares.
- Climate scientists inform us that if the carbon dioxide content of the atmosphere is to be below *twice* the pre-industrial level (many scientists argue it should be far lower than that), total emissions must be held below 9 billion tons per year. For 9 billion people, that is one ton per person. Present US and Australian per capita emission rates are around 16 tons per year, and for Australia, if land clearing is included, the figure is 27 tons!

These kinds of consideration provide impressive support for the conclusion that energy consumption and resource demand patterns in rich countries are already far beyond sustainable limits. Yet virtually all countries seek economic growth, and ignore any question of limits.

I have argued in a number of works that renewable energy sources and energy conservation cannot substitute for fossil-based energy supplies, and that technical advances, a "factor four" transition and "dematerialization" are not capable of solving the problem.[2]

THE "SIMPLER WAY" ALTERNATIVE

Given this "limits to growth" context, it should not be surprising that the discussion of desirable social forms leads to extremely radical conclusions. It is clear not only that a sustainable society cannot have a growth economy, but that consumption standards must become far lower than they are in rich countries at present. Nevertheless, advocates of the Simpler Way firmly believe that it could provide all with a higher *quality* of life than is typical of rich countries today.

The following is a broad outline of this vision. The intention is not to detail energy sources, production quantities and technologies utilized, but to sketch the kind of society we must shift towards if we are to solve global problems. Of course, this kind of alternate society would enable energy demand to be cut to far below present levels.

If the limits are as savage as they increasingly seem, then the essential and inescapable principles for a sustainable, alternate society must include:

1. **A simpler, non-affluent way of life.** We must aim at producing and consuming only as much as we need for comfortable and convenient living standards. We must phase out many entire industries. But living materially simply does not mean deprivation or hardship. There is no need to cut back on production of anything we need for a very comfortable, convenient and enjoyable way of life. The goal should be to be satisfied with what is sufficient.

2. **The development of many small-scale, highly self-sufficient local economies**. Most basic necessities should be produced very close to where we live. Declining energy supplies will prevent present levels of transport and packaging from being maintained, making economic decentralization a key requirement. We need to convert our neighborhoods, suburbs and towns into small, thriving

local economies which produce most of the goods and services they need, using local resources wherever possible.

Every suburb would have many small productive enterprises such as farms, dairies, local bakeries and potteries. Many existing economic entities would remain, but their operations would be decentralized as much as possible, with workers living close to their place of work, enabling most of us to get to work by bicycle or on foot. Many farms could be backyard and hobby businesses. A high proportion of our honey, eggs, crockery, vegetables, furniture, fruit, clothing, fish and poultry could come from very small, local family businesses and cooperatives. We would, however, retain some mass production facilities, but many items of general necessity such as furniture and crockery could in the main be produced through craft-working. It is far more satisfying to produce things using craftworking than in factories.

Market gardens could be located throughout suburbs and even cities – for example, on derelict factory sites and beside railway lines. Having food produced close to where people live would enable nutrients to be recycled back to the soil, through garbage (biogas) gas production units. This is essential for a sustainable society. Two of the most unsustainable aspects of our present agriculture are its heavy dependence on energy inputs and the fact that it takes nutrients from the soil and does not return them.

We should convert one house on each block into a neighborhood workshop. It would include a recycling store, meeting place, leisure resources, craft rooms, barter exchange and library. Because we will not need the car very much when we reduce and decentralize production, we could dig up many roads, thereby making perhaps one-third of a city's area available as communal property. We can plant community orchards and forests and put in community ponds for ducks and fish. Most of your neighborhood could become a permaculture jungle, an "edible landscape" crammed with long-lived, largely self-maintaining productive plants such as fruit and nut trees.

There would also be many varieties of animals living in our suburbs, including an integrated fish-farming industry. Communal woodlots, fruit trees, bamboo clumps, ponds and meadows would provide many community goods. Local supplies of clay could meet all crockery needs. Similarly, cabinet-making wood might come from local forestry, via one small neighborhood sawbench located in what used to be a car garage.

There is a surprising amount of land in cities that could be used to produce food and other materials. Firstly, there will be home gardening, the most efficient and productive way to provide food. Many flat rooftops can be gardened. Enabling the majority of persons to move from cities to country towns would make for more garden space in cities.

There is immense and largely untapped scope for deriving many materials from plants and other sources that exist or could be developed where we live: bark for tanning, dyes from plants, tar and resins from distilled flue gases, wool, wax, leather, feathers, paint from oil seeds like sunflowers, and many medicines from herbs. Small animals are easily kept within urban neighborhoods, and can yield many products including leather and fertilizer. Much of their feed could consist of recycled kitchen and garden waste. Timber would come from the woodlots and clay from the local pits. Many of these things would come from the commons we should develop in and around our settlements, including orchards, ponds, forests, fields, quarries, bamboo clumps, herb patches, and so on, which would be owned, operated and maintained by the community.

We could build most of our new housing ourselves, using earth and recycled materials, at a tiny fraction of present housing construction energy cost (and present cash cost).

We would also have decentralized, small-town banks run by elected boards, making our savings available for lending only to socially useful projects in our town or suburb. Local "business incubators" would help small firms to start up with low- or zero-interest loans where appropriate. We would then be in a position to create a dense network of many small firms that would enable unemployed people to start producing to meet those needs that are presently ignored. Because all our local small industries would be owned by people who live in our area, profits would not be siphoned out to distant shareholders but would be spent or reinvested in our area.

This would be a leisure-rich environment. Most suburbs at present are leisure deserts. The alternative neighborhood would be full of interesting things to do, common projects, animal husbandry, small firm activities, gardening, urban forestry and community workshops. Consequently, people would be less inclined to go away on weekends and holidays, again reducing national energy consumption.

Most of the things we need for everyday life in a sustainable society could be produced within a few kilometers of where we live; indeed, most needs would be satisfied by neighborhood production.

Some items, such as radios and stoves, could be produced in factories within 10 to 20 kilometers. Perhaps a small city might need one refrigerator manufacturer and repair center. Only a few specialty items might have to be transported hundreds of kilometers from large factories and very few would have to be imported from other countries – for example, high-tech medical equipment. Rational social and economic decisions would have to be made on the location of production facilities for goods that would be exported out of the local region, so that all towns and suburbs can earn a sufficient, small amount of export income to pay for their small import needs.

3. **More communal, cooperative and participatory practices.** The third necessary characteristic of a sustainable society is that it must be much more communal, cooperative and participatory than the society we know today. We must share more things. For example, we could have one stepladder in the neighborhood workshop, rather than one in most or many houses. We would give away surpluses. We would have voluntary community "working bees" to provide most child-minding, nursing, basic education and care of aged and handicapped people, as well as performing most of the functions that town and local councils presently carry out on our behalf, such as maintaining parks and streets.

The working bees and neighborhood committees would also maintain the many local commons, such as the orchards, woodlots, ponds, clay pits, workshops, windmills and other local renewable energy supply systems.

There would be a far greater sense of community than there is now. People would know each other and would constantly interact in community projects. One would certainly predict a huge decline in the incidence of loneliness, depression and similar social problems, and therefore in the cost of providing for people who have turned to drugs or crime, or who suffer stress, anxiety and depressive illness. It would be a much healthier and happier place to live, especially for young and old people. Markedly reduced rates of anomie, social stress and suicide can easily be predicted.

There would necessarily be a transition to a radically different form of government: to small-scale, local and participatory democracy. Most of our local policies and programs could be worked out by elected, unpaid committees, and we could all vote at general assemblies or town meetings on the important decisions concerning our small area. There could still be functions for state and national governments, but relatively few.

4. **Alternative technologies.** In some areas people could still use as much modern technology as they wished – in medicine and dentistry, metalworking, information technology, and so on – and much research could go into developing better technologies. However, in most areas use would be made of relatively simple, traditional and alternative technologies, because these have far lower resource and ecological impacts, and because they are more enjoyable and convivial. For example, most food will be produced by hand tools from home gardens, small local market gardens and permacultured "edible landscape" commons. These are the most enjoyable ways to produce the best food. Some farms will use some machinery, but on a small scale.

Water will mostly come from rooftops and pollution-free creeks and landscapes. Much manufacturing will be through crafts, hand-tools and small family firms and cooperatives. Many "services," such as the care of older people, will mostly be given informally and spontaneously within supportive communities, not via bureaucracies and professionals. We would research plant- and earth-based substitutes for some scarce minerals and chemicals. Many more tasks will be performed by human labor, as distinct from machines, such as cutting firewood and producing food, because this is more satisfying and because there will not be much energy available for running machines.

Although the Simpler Way seeks the simplest ways of doing things, it is not ideologically opposed to modern technology. Photovoltaic cells, for instance, are desirable, although they are technically complex. However, the Simpler Way notes that sophisticated modern technology is mostly unnecessary, and that technical progress is of little significance in solving the world's real problems or in providing a high quality of life to all. The key to these objectives is applying *simple* methods to meeting human needs while satisfying ecological requirements, which is not done in the present economy, or in our competitive, individualistic culture.

5. **An almost totally new economic system.** There is no chance whatsoever of making these changes while we retain the present consumer-capitalist economic system. This is the crucial implication

from the "limits to growth" literature. The major global problems we face are primarily due to this economic system. The new economy must be organized to meet the needs of people, the environment and social cohesion, with a minimum of resource use and energy consumption for a maximum quality of life. This is totally different from an economy driven by profit, market forces and growth.

The need for small, highly self-sufficient local economies, and for zero economic growth, has been noted. There will be relatively few big firms, little international trade, not much transporting of goods between regions, and very little, if any, role for transnational corporations and banks.

Market forces, free enterprise and the profit motive might be given a place in an acceptable alternative economy, but they could not be allowed to continue as major determinants of economic affairs. Basic economic priorities and structures must be planned for, provided and regulated according to what is socially desirable (democratically planned, mostly at the local level; not dictated by huge and distant bureaucracies). However, much of the economy might remain as a carefully regulated and monitored form of "private enterprise" carried on by small firms, households and cooperatives, so long as their goals were not profit-maximization and growth. There would have to be extensive discussion and referenda in deciding how to sympathetically phase out the many unnecessary and wasteful industries that now exist, and how to reclassify and redeploy their workers. Social machinery, especially the economy, is very complicated and problems can easily arise. A great deal of effort will have to go into operating, monitoring, debating and revising this machinery.

There would be a large and important non-monetary sector of the economy, including giving, mutual aid, volunteer work on committees, working bees, and the supply of free goods from local commons. Working bees' activity could be an effective way to pay "tax" – that is, to contribute to the maintenance of public facilities.

The new economy would probably have a relatively small cash sector, and would allow carefully regulated market forces to operate within it. Most of the important large enterprises, such as railways and steel, would probably be planned and run by collective or public agencies or firms, and at the local level would be operated by community cooperatives. Possibly the largest sector of the new economy would be run by community service cooperatives – the local energy supply or water supply cooperative "firms," for example.

Most of us would live well, with greatly reduced needs for cash income, because we would not need to buy very much. Consequently, many of us might work only one day a week for money, and spend the rest of the week work-playing around our neighborhoods in a wide variety of interesting and useful activities. There would be no unemployment and no poverty, as expressed in the ideal of Israeli Kibbutz settlements. We would have local work-coordination committees, which would make sure that all who wanted work had a share of the work that needed doing in the area. All people could make important economic contributions, even though some might have few educational qualifications or be mentally or physically handicapped, because there would be many simple but crucial jobs to be done in the gardens, workshops, forests and animal pens. All people could be fully active and valued participants in the economy.

There would be far less need for capital; capital-intensive factories and infrastructures such as roads, dams and power stations would be reduced, because the volume of production and transportation, and quantity of energy needed would be far less than they are now. There would be fewer types of products. For example, we might decide to have only a few types of radios, televisions, buses and cars, designed to last and to be repaired easily.

Few big firms and little heavy industry would be needed, because there would be much less production, especially of complex and sophisticated goods. Most items would be produced by small family firms and cooperatives in which people would invest their own savings, deriving modest, stable incomes. A few large firms would provide things like steel and railway equipment, and these should be run as public enterprises. Again, their control must be through open and participatory mechanisms, not necessarily by the state. There must be processes whereby all people can constantly monitor and evaluate the performance of public institutions and enterprises.

The focus in the above account has been on the neighborhood and suburban economy. Beyond these local economies there would still be regional, national and international economies, but their activity levels would be far lower than at present.

6. **New values.** Obviously, the Simpler Way will not be taken unless there is change from the presently dominant values and habits.

There must be a much more collective, less individualistic social philosophy and outlook; a more cooperative and less competitive attitude; a more participatory and socially responsible orientation; and above all, much greater willingness to be satisfied with less, and by what is simple but sufficient.

These are the biggest difficulties facing the transition to a sustainable society. However, it is important to recognize that the society we have now forces us to compete against each other, for example for jobs, and that people now consume mainly because few other sources of satisfaction or meaning are open to them in consumer-capitalist society.

On the other hand, the Simpler Way offers many satisfactions and rewards: if people can be helped to see this they will be more likely to move away from consumer society. Consider, for example, having far more time outside the economic nexus, and having to work for money only one or two days a week, living in a rich, varied and supportive community, with interesting, enjoyable, varied and worthwhile work to do, contributing to the governance of one's community, participating in meetings and decision-making for the good of all. Consider also having much more time for learning and practicing arts and crafts, for personal development and for community development, participating in many local festivals and celebrations, running a productive, efficient and highly self-sufficient household and garden. Consider living in a leisure-rich environment, being secure, not having to worry about unemployment or being lonely, not worrying about how you will cope if you are ill or when you are aged. Above all, there is benefit in knowing that you are no longer part of the global problem, because you are living in ways that are sustainable.

Compared with people in the consumer society of today we would be very poor, wearing old clothes, living in small, sometimes mud-brick houses, and earning very low cash incomes. However, the Simpler Way makes possible a much better quality of life than most people in rich countries experience at present.

There is nothing backward or primitive about the Simpler Way. We would have all the high-tech and modern tools that make sense, such as in medicine, renewable energy technology, public transport and simpler, energy-saving household appliances. We could still have smaller but effective national systems for many things, such as railways and telecommunications. We would also have far *more* resources for science and research, and for education and the arts, than we do now, because we would have liberated those resources presently being wasted, for example in the production of unnecessary items, including arms.

CONCLUSION

The core claim made here is that only if we move, both in rich and poor countries, to something like this vision of the Simpler Way can we expect to achieve a just and sustainable global situation. Only by instituting materially simple, self-sufficient and cooperative practices within a new, zero-growth economy, can we hope for a high quality of life at much lower levels of energy use.

There is now a global alternative society movement gathering momentum throughout the world, in which many small groups are actually building settlements more or less along the lines outlined above.[3] The fate of the planet depends on how successful this movement will be in creating demonstration settlements, and proving their feasibility before the problems in rich countries become so acute that a reasoned, ordered and sensible transition becomes impossible.

23
Musing Along

Andrew McKillop

The Muses, like their near-opposites Spites were Greek mythical forces or categories of beings, each with a specific mythological explanation of how they came about and what they represented. A "muse" today is something akin to inspirational guidance, and spite is a mean-minded desire for revenge people have when things do not go their way. However, when these two words are used with capital letters in the Greek mythological sense, they obliquely refer to much more critical events and forces: to life-and-death choices in land use, food production, population movements, and to basic human survival in the face of what appears to be natural catastrophe, but in fact is often human-induced.

According to Greek myth, the Muses were nine frolicking, uninhibited junior goddesses on Mount Olympus who were tamed by Apollo. They became almost mirror opposites of their former selves, and inspired restraint, moderation, care and forward-thinking behavior in those gods, half-gods and mortals who chose to take their advice. Sometimes the Muses acted as three groups of three – for example, in the complex, chaotic, much-modified Oedipus myth involving his slaying of the Sphinx, as well as his apparently innocent father-slaying. This was followed by his equally innocent but disastrous marriage to his mother, leading her to suicide when she found out the true identity of her new husband, and to the banishment and death of Oedipus. Sometimes all nine Muses would take the scene – for example, in the series of myths that include the Labors of Heracles, these trials only being consented to by Heracles while he was recovering from madness, visited on him by Hera, the supreme female god, perhaps out of jealousy or spite.

The central roles of the Muses, towards the end of Ancient Greek civilization and culture (which ended at about 400BC), shrank down to the "woman's roles" of prophecy and healing. While formerly they had, with reformed Apollo, insisted on moderation and self-knowledge being the best foundations for the behavior and action of mortals and gods, in their later role they became humdrum suppliers of forecasting tools and methods. No longer giving moral advice, they advised on how to interpret pebbles sinking in bowls of water, and how to throw divinatory knucklebones for gauging what action to take. The Thriae or Triple-Muse is credited with teaching Hermes, a god, how to use knucklebone forecasting, this method then being taken back to Olympus for all the gods and goddesses to use and approve. As Robert Graves and others[1] have suggested, these powers and roles of Greek mythical entities, the Muses in this case, changed in a radical but seemingly haphazard way, through a long period we can place at about 3000–400BC, because of social, cultural, economic *and environmental* changes. Many of these latter changes were drastic. The two periods of greatest change could be dated at about 2500BC, when all the cultures of the East Mediterranean had become patrilineal and patriarchal (after having been matrilineal and matrianehal); and then somewhat later, probably from about 1500BC, when large population movements and even greater environmental changes took place.

Perhaps no single myth better incorporates these changes than the Demeter or "corn goddess" myth. Demeter never married, but in no way was inexperienced, having her first sexual experience with her father, Zeus. After her later orgiastic experience with Iasius in a triple-ploughed field (ploughed three times in a year), she progressively retreated from sexual contact with male gods. When she lost her daughter Core, abducted by Hades, Demeter pined for nine days and nights, then ragingly condemned the world to barren sterility – even plants ceased to grow. A complex deal was struck by Zeus to save the world's food production. Demeter's daughter would remain the companion (or "wife") of Hades, and be called Persephone in her underworld role, for six months a year, returning to the upper world to live as Core for the other six. While Persephone, she had the role of deciding, by pulling a single hair from her head, who would live or die. In her underground semester of winter, mythological rites required that Core, as a corn doll, be buried, before her return to the upper world in the spring.

Demeter had at least one other identity – Nicippe, the protector of forests. Almost certainly in early Greek myth Nicippe fiercely protected forested land, and was at least as powerful as Demeter. Even when Demeter became the goddess of agriculture, the residual forest-protecting role was held by Nicippe. This declining role for forested land and female food-gathering (with increasing population and the growth of agriculture) is explained by Nicippe's clash with Triops, whose son disrespectfully cut down a grove of small, sacred oak trees, for timber to construct a banqueting hall. He ignored her call to desist, so Nicippe condemned him to perpetual hunger, as Midas was condemned, though later reprieved by Dionysus, to have anything he touched turn to gold, of which Midas had wanted unlimited quantities for his trading and commercial pursuits. Triops' son was not celebrating trade surplus, but agricultural surplus. One condition for that was to cut down forest and triple-plough the resulting, bare land. This of course changed hydrographic and even local climatic conditions, and increased erosion, leading to the silting of bodies of water. While triple-ploughing of deforested land and tri directional ploughing of the resulting fields enabled three harvests to be mapped in one year, this practice was condemned by the waning, and finally ineffectual Nicippe myth. Demeter, however, as a purely *agricultural* goddess, became ever more powerful with increasing agricultural surpluses. In her waning and declining role as Nicippe, one version has it that Erysichthon, the son of Tiops, commits the fault of deforestation. Even his name gives a clue to Nicippe's injunction on over-ploughing: Erysichthon means "he who rips the earth."

Many subtle references to culture change, enabled or forced by population growth and environment modification, are included in this and other Greek myths. The coming of agriculture – both a result and cause of increasing human numbers – ended female domination of society. No longer was gathering, by women, the main source of food. Consequently, female deities, spirits and forces progressively shrank, even physically, and certainly in status – for example, in abduction and demotion myths, where female deities submit to male dominance. Likewise older deities and powers, dominated by female entities and natural forces (including animals) give way to exclusively human deities and powers, mostly masculine. Previous and of course rudimentary population control methods were replaced. It is interesting to consider which were the more barbaric – the old or the new. However, simply because of numbers, the later male-dominated "population control" methods were unquestionably more bloody but (also because of numbers) *less* effective. Lifelong chastity, infanticide and the ritual slaying of young people at puberty, used in the earlier matriarchal and matrilineal phases of societies, were substituted by homosexuality, war, the killing of captives and ritual destruction of agricultural surplus under male deities and patriarchal societies. This latter practice, the ritual destruction of agricultural surpluses (both animal and crop) was mythologized in the changeover from human blood sacrifice to that of oxen, cattle, and then horses, their blood being spread over ploughed fields in the belief that it increased the fertility of the land. Already the notion of fertilizers can be seen in such actions, the blood having a physical utility as well as its previous ritual power.

Like many older civilizations and cultures, Greek myth records the impacts of agriculture and coastal works like harbor building and dredging, along with city construction which, by about 1000BC, could extend to dozens of square kilometers. Thus, from the Yorubas of West Africa to the Gaels of the Hebrides, the ancient Greeks, Egyptians, Jews, Palestinians and others refer to sunken or lost lands, even the magical continent of Atlantis. In Greek myth, this was the home of Atlas, the Titan and brother of Prometheus and Epimetheus, and of his peoples, who were also the inventors of the earliest alphabets. Atlantis was remarkable for its huge ports, irrigation and canalization works, roads, and agricultural endeavors. However, the people became too greedy, vain and cruel to their defeated enemies, and insulted or ignored the gods. Zeus decreed the destruction – by flooding – of Atlantis. Atlas himself, and certain other Titans, escaped and plotted revenge, together with Cronus and other "antique" female-dominated deities – Cronus being present in the very first or Pelasgian (pre-Hellenic) myths. Pelasgus also means "stork," and Cronus is often portrayed as a giant crow. While these details can be amusing or confusing, the essence of at least three myths – of the Deucalion Flood, of Dionysus (the inventor of wine), and of the Giants' Revolt or War of the Gods (between old and new gods) – reflects large-scale land-use modification, agriculture, urbanization and changing hydrographic regimes due to human activities. The biblical myth of Noah is nothing but a later retelling of a flood myth, even down to the conservation of plant and animal species for future production, and to the detail of Noah in Hebrew myth being the inventor of wine, the production of alcohol enabling increased conservation of foods, as well as many other technical processes, and of course alcoholic excess.

Prometheus was the creator of mortals, a project that he had started by saving earlier Pelasgian mortals, descended from Deucalion, who grew like vine or ivy leaves. (Yet earlier versions had grown from snake teeth, or in Irish Gaelic myth were "mushroom men.") Deucalion had received the vines, or ivy, or snake's teeth, or mushrooms from a Moon goddess, or from a magical deer or wolf. These earlier Pelasgian mortals were unfit to exist, according to Zeus, because of their fornication, cruelty and cannibalism, and their great aggressiveness amongst themselves, reproducing in huge numbers only to destroy each other for trivial causes. For this reason he decided to drown them all in the Deucalion flood, but Prometheus – while siding with Zeus in his war on the Titans and destruction of Atlantis – warned some Pelasgians of the flood, allowing them to escape it on various mountain tops, and of course in arks containing seed plants and domestic animals.

Prometheus above all strove to give mortals a second chance. Knowing their weaknesses – the "Spites," including Vanity, Violence, Useless (or Heroic) Labor, Delusive Hope, Sex without Love, Rhetorical Denial, Insanity and other recognizably human attributes – he painstakingly rounded up the Spites and trapped them in a huge amphora jug. Zeus was of course aware of this, knew where the jug was hidden, and still nurtured anti-human sentiments. When Zeus learned that Prometheus and his brother, Epimetheus, were intent on re-launching the project of allowing mortals to divide and multiply, without any doubt copying the excess of Atlas and using his works (including alphabets), Zeus decided to abort the project. He did so by creating Pandora, a "clay woman" of incredible beauty, and, after equipping her with the jug, gave her to Epimetheus, expecting him to accept this poisoned gift, together with the amphora jug (later called a "box"). Epimetheus however, had, been warned by his brother to accept nothing from Zeus – even Pandora, barely clad in all the finery, and skilled in all the seductive artifices, of female gods. So Epimetheus politely refused the gift, enraging Zeus so that he had Prometheus tortured, day and night, attached to a post in the Caucasian mountains, for nine years, his liver being torn out each day by wild birds and beasts, but growing back each night. As might be expected, Epimetheus decided finally to accept Pandora, to liberate his brother, and Pandora lost no time in opening the fateful jug. Only because of the Spite called Delusive Hope did the brothers not commit suicide, goes the tale.

Consequently, human beings were struck by the Spites, and equally by an obsessive desire to change the face of the planet. Among the consequences were and are floods, famines and migrations – preceded or followed by war and infanticide. To the Ancient Greeks, these were inevitable. These and other myths from the period of about 3000–1000BC, in a wide region of the planet (perhaps including

Central America), all trace the development and growth of food production and urbanization. Ancient Greek civilization collapsed, or was "decanted," and transferred into Roman civilization in the period of about 500–200BC; the Romans having neither the time nor creative energy to initiate or develop any particular system of myths, because their civilization of no more than 1,000 years in duration was based, from the start, solely on expansion.

Our fossil-energy civilization, excluding the coal-based phase, can be given a total span or "useful lifetime" of about 1850–2035. Its cultural and mythical "creation" is limited to the instant myths accompanying TV celebrities or politicians, and to a mass cult of denial regarding environmental and planetary limits. Its enduring legacy will include suboptimal human population numbers after "adjustment," or die-off, following the Final Energy Crisis.

Notes on Contributors

Colin J. Campbell has worked as an oil exploration geologist in many parts of the world, including Australia, Trinidad, Colombia, and New Guinea. In 1968, he joined Amoco in New York to work on global new ventures and resource assessment. In 1984, he became Executive Vice-President of Fina in Norway. Since 1989 he has provided independent advice to governments and major oil companies. He is the author of five books and has published many articles and has lectured widely. Most recently he founded ASPO, an international Association for the Study of Peak Oil.

André Crouzet is the conference and publicity coordinator for the anti-nuclear organization Sortir du Nucléaire.

Jacob Lund Fisker was born in Denmark. He obtained his Ph.D. in theoretical physics at the University of Basel, Switzerland, in 2004, and since then he has worked as a research associate at the Joint Institute for Nuclear Astrophysics at the University of Notre Dame in Indiana, USA, on Hydrodynamical simulations of white dwarf accretion disks. His interest in Peak Oil stems from his observation that the whole of our civilization seems dependent on oil.

Edward R.D. Goldsmith is an editor, author, lecturer, and campaigner. He founded *The Ecologist* magazine in 1969. Awarded the Honorary Right Livelihood Award (the Alternative Nobel Prize) in 1991, he is a critic of globalization, and governments and corporate damage to the ecological foundations of life on earth.

Mark Jones was an author and novelist with a longtime interest in the political economy of oil. He lived and worked in the Soviet Union and in post-Soviet Russia, working for a time with the *New Statesman*, whilst researching archives newly opened under Gorbachev's policy of Glasnost. He founded the Conference of Socialist Economists and their journal, *Capital and Class*, in the mid 1970s and in the 1980s he was remanded in custody for publishing Leon Trotsky's writings. He sustained a deep interest and sympathy for communism, and was concerned by the limits to growth and believed that the US economy had an unsustainable but unstoppable need for oil. His novel *Black Lightning*, about the fall of the

Soviet Union, was published in 1995 by Victor Gollancz. He died in England in 2003.

Seppo A. Korpela is a professor of mechanical engineering at the Ohio State University, where he has been teaching since 1972. His teaching and research interests are in applications of heat transfer. He has contributed to optimizing the gap width of double and triple pane windows, and developed a theory for heat transfer by conduction and radiation through fiberglass insulation. He is a member of the Green Building Forum of Columbus, Ohio; a group formed to promote sustainable building practices.

William Ross McCluney is Principal Research Scientist at the Florida Solar Energy Center, a research institute in Cocoa of the University of Central Florida in Orlando, Florida, USA. Subsequent to taking his current position in 1976, he taught a graduate course in optical oceanography at Florida Institute of Technology as an adjunct professor and supervised student research work at Florida Tech in oceanography and physics. In 2003–04, Dr. McCluney taught a UCF course on Philosophy, Religion, and the Environment. In 2004 he gave the plenary session lecture to Canada's University of Waterloo Solar Energy Society on justifications for energy efficient and renewable energy technologies. He is the author of *The Environmental Destruction of South Florida* (University of Miami Press, 1971), the popular textbook *Introduction to Radiometry and Photometry* (Artech House of Boston, MA, 1994), *Getting to the Source: Readings on Sustainable Values*, and *Humanity's Environmental Future*, the latter two published by SunPine Press, Cape Canaveral, FL, 2004.

Andrew McKillop was born in Bern, Switzerland and is a writer and consultant on oil- and energy-economics, with wide international experience. Since 1975 he has worked in national and international energy, economic, scientific, and administrative organizations in Europe, Asia, the Middle East, and North America. He is also a translator of English into French with a background in financial, technical, conference and policy documents. He has formal qualifications in Economics, Environment and Energy studies, and has language qualifications. He is a founding member of the Asian chapter of the International Association of Energy Economics.

Sheila Newman is a bi-lingual independent sociological researcher in Australia. She writes on environment, population, energy and land-use planning and inheritance systems. She is interested in comparing the role of incest avoidance in population spacing algorithms in other species with inheritance factors in human land-use allocation systems. As an artist, working mainly in electronic media and animation, her concerns are the destruction of urban bushland and wildlife in Australia, and soil depletion and over-development of water catchments. She makes digital movies with Quark and Neutrino productions about how land-speculation entwines with population pressure and corporatization of government inevitably leads to science being sidelined, democracy diminished, and ecological disintegration.

Mark Saint Aroman has managed the activist network for the anti-nuclear organization Sortir du Nucléaire since it began in 1997. He has a background as an electrical technician in radiology and publishes technical information about nuclear energy.

Ted Trainer is a lecturer in the School of Social Work, University of New South Wales. His main interests have been global problems, sustainability issues, radical critiques of the economy, alternative social forms and the transition to them. He has written numerous books and articles on these topics, including: *The Conserver Society: Alternatives for Sustainability* (London: Zed, 1995); *Saving the Environment: What It Will Take* (Sydney: University of NSW Press, 1998), and *What Should We Do* (forthcoming). He is also developing Pigface Point, an alternative lifestyle educational site near Sydney, and a website, http://www.arts. unsw.edu.au/tsw/

Gregson Vaux graduated *magna cum laude* in physics from the University of Pittsburgh and obtained his graduate degree in civil and environmental engineering from Carnegie Mellon University. He previously taught at the School of Petroleum Engineering at Kuwait University and currently is employed in the power and energy industry as an engineer and analyst.

Notes

PART I

Chapter I

1. Hubbert, M. King, "Techniques and Prediction as Applied to the Production of Oil and Gas," *Proceedings of a Symposium on Oil and Gas Modeling*, pp. 16–142 (1982).
2. Campbell, C.J. and Laherrère, J., "The End of Cheap Oil," *Scientific American* (March 1998).
3. Campbell, C.J., *The Coming Oil Crisis*, Multi-Science Publishing Co. & Petroconsultants S.A. (1998).
4. Campbell, C.J., "Better Understanding Urged for Rapidly Depleting Oil Reserves," *Oil & Gas Journal* (April 7, 1997).
5. Campbell, C.J., Monthly Newsletter of the Association for the Study of Peak Oil, www.energikrise.de (November 11, 2002).
6. Laherrère, J., "World Oil Supply, What Goes Up Must Come Down, But When Will it Peak?" *Oil & Gas Journal* (February 1, 1999).
7. Ivanhoe, L.F., Director, M. King Hubbert Center for Petroleum Supply Studies, www.hubbert.mines.edu (November 1, 2002).
8. Duncan, R.C. and Youngquist, W., "Encircling the Peak of World Oil Production," *Natural Resources Research* [vol. 8, no. 3] pp. 219–31 (1999).
9. Youngquist, W., *Geodestinies*, National Book Company, Oregon (1997).
10. Bakhtiari, A.M.S. and Shahbudaghlou, F., "IEA, OPEC Supply Forecast Challenged," *Oil & Gas Journal* (April 30, 2001); Bakhtiari, A.M.S., "OPEC Capacity Potential to Meet Projected Demand Not Likely to Materialize," *Oil & Gas Journal* (July 9, 2001).
11. Deffeyes, K.S., *Hubbert's Peak*, Princeton University Press (2001).
12. Hubbert, "Techniques and Prediction."
13. United States Geological Survey, www.greenwood.cr.usgs.gov/energy/WorldEnergy/OF99-50Z (November 1, 2002).
14. Campbell, *Coming Oil Crisis*.
15. United States Geological Survey, www.greenwood.cr.usgs.gov/energy/WorldEnergy/OF99-50Z (November 1, 2002).
16. Hubbert, "Techniques and Prediction."
17. Campbell, C.J., "USGS, Global Petroleum Reserves – a View to the Future," www.oilcrisis.com/news/article.asp?id=3659.
18. "Worldwide Reserves Increase as Production Holds Steady," *Oil & Gas Journal* (December 23, 2002).
19. Mineral Management Service, "MMS Updates Oil & Gas Production Rate Projections to 2006, Steep Rise in Oil," www.gomr.mms.gov/homepg/whatsnew/newsreal/020610.html (June 1, 2002).
20. Bakhtiari and Shahbudaghlou, "OPEC Supply Forecast Challenged."

21. McCabe, P., "Energy Resources – Cornucopia or an Empty Barrel?" *Bulletin, American Association of Petroleum Geologists* [vol. 82] pp. 2110–34 (1998).

22. Laherrère, J., "Estimates of Oil Reserves" (Fig. 90), EMF/IEA/IEW Meeting, IIAW Austria (June 19, 2001).

23. Bartlett, A., "An Analysis of US and World Oil Production Patterns Using Hubbert-style Curves," *Mathematical Geology* [vol. 32, no. 1] (2000).

24. Deffeyes, *Hubbert's Peak*.

Chapter 3

1. The Hadley Centre, *Modeling Climate Change, 1860–2050*, Met Office (February 1995).

2. Bunyard, Peter, "Misreading the Models, Danger of Underestimating Climate Change," *The Ecologist* [vol. 29, no. 2] p. 75 (March/April 1999).

3. IPCC, "Third Assessment Report," Cambridge University Press (1995).

4. Bunyard, Peter. "How Global Warming Could Cause Northern Europe to Freeze," *The Ecologist* [vol. 29, no. 2] (March/April 1999). http://newfirst. search.oclc.org (21 May 2002).

5. Bunyard, Peter, "Industrial Agriculture – Driving Climate Change," *The Ecologist* [vol. 26, no. 6] pp. 290–8 (November/December 1996).

6. *Ibid.*

7. Moser, A. et al., "Methane and Nitrous Oxide Fumes in Native Fertilized and Cultivated Grassland," *Nature* [vol. 350] (March 1991).

8. Tebruegge, F., *No-Tillage Visions – Protection of Soil, Water and Climate*, Institute for Agricultural Engineering, Justus-Liebig University, Germany (2000).

9. Smith, P. Powlson, D.S. Glendenning, A.J. and Smith, J.U. "Preliminary Estimates of Potential Carbon Migration in European Soils Through No-Till Farming," *Global Change Biology* 4, quoted by Smith, Corrinne in *L'Ecologiste* [vol. 3, no. 7] (June 2002).

10. Pretty, J. and Ball, A., "Agricultural Influences on Carbon Emissions and Sequestration: A Review of Evidence and the Emerging Trading Options", Centre for Environment and Society and Department of Biological Sciences, University of Essex, UK, Centre for Environment and Society Occasional Paper, 2001–03, University of Essex, March 2001.

11. Payer, C., *The World Bank, A Critical Analysis*, Monthly Review Press (1982).

12. *Ibid.*

13. World Bank, "Accelerated Development in Sub-Saharan Agriculture", Washington (1981).

14. *Ibid.*

15. McKenney, Jason, "Artificial Fertilizing", in Andrew Kimbrell, *The Fatal Harvest Reader*, Island Press, Washington DC (2002), p. 182.

16. *Ibid.*

17. Leach, Gerald, "The Coming Decline of Oil," *The Pacific Ecologist*, New Zealand, pp. 34–6 (Summer 2002–03).

18. FAO, *Sequestration de carbonne terrestre pour une meilleure gestion du sol. Rapport de la FAO 2001*, quoted by Smith, Corrinne in *L'Ecologiste* [vol. 3, no. 7] (June 2002).

19. Carson, R., *Silent Spring*, Hamish Hamilton, London (1963), p. 48.

20. Pretty, J., "Regenerating Agriculture Policies and Practice for Sustainability and Self-Reliance", Earthscan, London (1995), p. 121.

21. *Ibid.*

22. Bonsu, M., "Organic Residues for Less Erosion and More Grain in Ghana," (1983) in M. el Swaify et al., *Soil Erosion and Conservation*, Soil Conservation Service, Ankery – Iowa; quoted by Pretty, J., "Regenerating Agriculture Policies."

23. FAO, *Sequestration*.

24. IPCC, "Third Assessment Report."

25. *Ibid.*

26. Pretty and Ball, "Agricultural Influences."

27. Farrell, John C., "Agroforestry Systems," from Altieri, M., *Agro Ecology, The Scientific Basis of Alternative Agriculture*, University of California, Berkeley (1985).

28. Pimental, D., "Global Climate Change and Agriculture," College of Agriculture and Life Sciences, Cornell University (1998).

29. Rosenzweig, C. and Hillel, D., *Climate Change and the Global Harvest, Potential Impacts of the Greenhouse Effect on Agriculture*, Oxford University Press, p. 29 (1998) [quoted by Bunyard, P. in "A Hungrier World," "The Ecologist Special Issue – Climate Crisis" (vol. 29, no. 2) p. 87 (1998)].

30. Briscoe, M., "Water the Overtapped Resource," in Andrew Kimbrell "The Fatal Harvest Reader," Island Press, Washington, DC (2002), p. 182.

31. *Ibid.*, p. 190.

32. *Ibid.*, p. 184.

33. Rosenzweig and Hillel, *Climate Change and the Global Harvest* in Bunyard, "A Hungrier World," p. 87.

34. Shiva, V., *Water Wars*, India Research Press, New Delhi (2002).

35. Goldsmith, E. and Hildyard, N., *The Social and Environmental Effects of Large Dams*, Sierra Club Books, San Francisco (1984).

36. Nellithanam, J. and R., "Return of the Native Seeds," *The Ecologist* [vol. 28, no. 1] pp. 29–33 (January/February 1998).

37. Agarwal, A. and Narain, S., "Traditional Systems of Water-Harvesting and Agroforestry" from *Geeti Sen Indigenous Vision – People of India Attitudes to the Environment, India*, International Centre, Sage Publication, New Delhi (1992).

38. *Ibid.*

39. Lange, T. and Hines, C., "The New Protectionism, Protecting the Future Against Free Trade," *Earthscan* (1993) [quoted by Jones, A. *Eating Oil, Food Supply in a Changing Climate*, Sustain and Elm Farm Research Centre (2001)].

40. Sims, A. et al., *Collision Course, Free Trade's Free Ride on the Global Economy*, New Economics Foundation (2000) [quoted by Jones, *Eating Oil*, notes 47; 49].

41. Jones, *Eating Oil*, p. 10.

42. Hewett, C., "Clean Air, Green Futures" [quoted by Jones, *Eating Oil*, p. 29].
43. Sewill, B., "Tax Free Australia," Aviation Environment Federation (December 2000) [quoted by Jones, *Eating Oil*, p. 30].
44. Jones, *Eating Oil*, p. 10.
45. Madeley, J., "Does Economic Development Feed People?" *The Ecologist* [vol. 15, no. 1/2] (1985).
46. *The Ecologist*, Special Issue on the FAO [vol. 21, no. 2] (March/April 1991).
47. "FAO Report on 1980 World Census of Agriculture," [census quoted by Shiva, V. in *Yoked to Death, Globalization and Corporate Control of Agriculture*, p. 13].
48. Mellanby, K., *How to Feed Britain* (1992).
49. Rossett, P., "What's So Beautiful About Small?" *Food for Life* (Summer 2000).
50. Lutzenberger, J., Personal communication.
51. Scott, J., "The Subsistence Ethic," *The New Ecologist* [no. 3] (May/June 1978).
52. Richards, P., *Cultivation, Knowledge and Performance*, London (1986).
53. Steudler, P.A.; Bowden, R.D. et al., "Influence of Nitrogen Fertilization on Methane Uptake in Temperate Forest Soils," *Nature* [vol. 341] pp. 314–15 (September 1989).
54. Ashton, J. and Laura, R., *The Perils of Progress*, Zed Books, London (1999).
55. Pearce, F., "Sea Life Sickened by Urban Pollution," *New Scientist* p. 4 (June 17, 1995) [quoted by Ashton and Laura, *The Perils of Progress*, note 61, p. 38].
56. Ashton and Laura, *The Perils of Progress*.
57. Junge, H.D. and Handke, S., "Nitrate in Vegetables – Unavoidable Risk?" *Indutrielle Obst – und Gemusewerwertung* [vol. 71, no. 8] pp. 346–8 (1987).
58. "Fertilizer Risks in Developing Countries," *Nature*, pp. 207–8 (July 21, 1988).
59. *Organic Farming Food Quality and Human Health, A Review of Evidence*, Soil Association (2001).
60. Goldsmith, Edward, "How to feed people under a regime of Climate Change", Note 9, "In Sri Lanka a traditional farmer (Mudiyense Tennakoon) told me that Sri Lankan farmers used to have no difficulty in keeping traditional strains of rice for 3–4 years, however the hybrid varieties using artificial fertiliser get mouldy in 3 months. [59]" www.edwardgoldsmith.com/page97.html.

Chapter 4

1. This idea forms the basis of what was later called the "Zeroth Law" of thermodynamics. Succinctly stated this law says that if an object, A, is in thermal equilibrium with objects B and C, then B and C are in thermal equilibrium too.
2. A few grams of matter contain about 10^{24} or one trillion trillion particles.
3. This theorem was initially proved by Boltzmann in 1872.
4. Waste heat can be observed as vapor leaving the cooling towers of electric power plants, the chimney of a steam locomotive, the exhaust pipe of a car, etc.
5. See www.eia.doe.gov/neic/brochure/infocard00.htm.

PART II

Chapter 6

1. US EIA; www.asponews.org/plots.php.
2. www.darksky.org; www.lightpollution.it.
3. The importance of Russian arms and war loans to Angola through the 26-year civil war (where the US backed Jonas Savimbi's UNITA forces) is that these are secured by Angola's *future* oil production, as are the loans provided by Swiss and French banks to the winning MPLA regime. Consequently, there is nothing left over for development, and Angola's newfound freedom will be "celebrated" in perfect poverty. See also, Hodges, T., *Angola: from Afro-Stalinism to Petro-Diamond Capitalism*, Indiana University Press (2001).
4. Eberstadt, N., "The Future of AIDS," "Foreign Affairs" Council on Foreign Relations (November/December 2002). This article focuses on the coming Eurasian pandemic (Russia, India and China) but notes that in Africa, in less than ten years, the number of HIV infected persons increased by at least 250 per cent, and that by late 2002 some 25 million persons have likely died from AIDS or AIDS-related disease in Africa. Many African countries have 15 per cent of their population HIV positive. By 2020, this could increase to 25 per cent to 35 per cent. (Data from projects funded by the Global Fund to Fight AIDS, Tuberculosis and Malaria, Geneva; and *Oeuvre médical au Congo*, Mr. Honoré Nkusu Zinkatu Konda, OMECO.)

Chapter 8

1. Leclercq, J., *L'ère nucléaire*, Editions Hachette, Paris (1997).
2. *Atomes Crochus* (film) ["We're on the same wavelength"] (1999).
3. Leclercq, *L'ère nucléaire*.
4. Report to Prime Minister July 2000, Annex 1, scenario S7.
5. M. Stoffaes is EDF's forward planning and international relations director.
6. "Financial Economics Association Review" [*Revue de l'Association d'Economie Financière*] [no. 66] Johannesburg (2002).
7. "EastWest Challenges; Energy and Security in the Caucasus and Central Asia", Report from the Conference held in Stockholm September 3–4, 1998, www.ui.se/news.htm#publika.
8. By way of the French "Bataille Law" (named for parliamentarian presenting the law for vote).
9. *Libération*, Paris (December 4, 2002).
10. Report no. 3415 to National Assembly, p. 51 (report on NRSE).
11. AMPERE Committee, National Assembly, www.ampereeurope.org/start.htm and www.ampereeurope.org/commit.htm.
12. The OECD's NEA (pro-nuclear agency), based on a study by the US Global Foundation, Inc., projects that in 2100 the capacity of nuclear reactors will be increased 18-fold from today, producing 44,000TWh per year.

PART III

Chapter 10

1. FIA website, www.ecosur.mx./scolel (2002).
2. International Carbon Sequestration Common Air Project website (2002).
3. *Ibid.*
4. CNRS Aboretum, Alsace, France (2001).
5. Pimental, D.; Bailey, O.; Kim, P.; Mullaney, E.; Calabrese, J.; Walman, L.; Nelson, F. and Yao, X., *Will Limits of the Earth's Resources Control Human Numbers?* College of Agriculture and Life Sciences, Cornell University, NY (February 1999) www.dieoff.com/page174.htm.

Chapter 11

1. Campbell, C.J., "Petroleum and People," *Population and Environment* [vol. 24, no. 2] pp. 193–207 (November 2002).
2. FAO (United Nations Food and Agriculture Organisation), "Progress in Reducing Hunger Has Virtually Halted – Chronic Hunger Kills Millions Each Year – Especially Children" (Rome 2002).
3. Pimental, D.; Bailey, O.; Kim, P.; Mullaney, E.; Calabrese, J.; Walman, L.; Nelson, F. and Yao, X., *Will Limits of the Earth's Resources Control Human Numbers?* College of Agriculture and Life Sciences, Cornell University, NY (February 1999) www.dieoff.com/page174.htm.
4. Steinhart, C. and Steinhart, J., *The Fires of Culture, Energy Yesterday and Tomorrow*, Wadsworth Publishing, Duxbury Press (1974).
5. Klare, Michael, *Resource Wars, The New Landscape of Global Conflict*, Henry Holt, New York (2002).
6. One giga-ton oil equivalent, or Gtoe, is equal to 41.868 EJ and 39.68 quadrillion (peta) BTUs.
7. Baron, S., *Mechanical Engineering* [vol. 103] p. 35 (1981).
8. Knudson, P., "New Mexico Monthly Bulletin" [vol. 3] p. 16 (1978).
9. Odum, H.T., Mitsch, W.J. et al., "Net energy analysis of alternatives for the United States," in *Hearings before the Subcommittee on Energy and Power, Committee on Interstate and Foreign Commerce*, US House of Representatives, March 25–26, 1976, Washington DC pp. 253–302E.
10. Baron, *Mechanical Engineering*.
11. "Concerted Action on Offshore Wind Energy in Europe CA-OWEE," *Renewable Energy World* [vol. 5] p. 28 (2002).
12. "Ocean Current Technical FAQ," Practical Ocean Energy Management Systems, Inc. (2000) www.poemsinc.org/currentFAQ.html.
13. Stewart, Harris B., Jr., "Proceedings of the MacArthur Workshop on the Feasibility of Extracting Usable Energy from the Florida Current," Palm Beach Shores, Florida (February 27–March 1, 1974).
14. *Ibid.*
15. "Ocean Tropics" Office of Energy Efficiency and Renewal Energy, US Department of Energy (December 18, 2002) www.eren.doe.gov/RE/ocean.html.

16. "Geothermal Energy Facts," Geothermal Education Office (March 14, 2002) www.geothermal.marin.org/geonergy.html#future.
17. "DTI Sustainable Energy Programmes – Introduction," UK Department of Trade and Industry (2002) www.dti.gov.uk/renewable.
18. "Geothermal Energy Facts."
19. McCluney, R., "Just Enough," *Solar Today*, p. 7 (1992).
20. Bartlett, Albert A., "Reflections on Sustainability, Population growth, and the Environment," *Renewable Resources Journal* Natural Resources Foundation, Bethesda, Maryland [vol. 15, no. 4] pp. 6–23 (winter 1997/8).

Chapter 12

1. Wackernagel, M. and Rees, W., *Our Ecological Footprint – Reducing Human Impact on the Earth*, New Society Publishers, Canada and US (1996).
2. Schulz, N.; Wackernagel, M.; Deumling, D.; Callejas-Linares, A.; Jenkins, M.; Kapos, V.; Monfreda, C.; Lohl, J.; Myers, N.; Norgarrd, R. and Randers, J., "Tracking the Ecological Overshoot of the Human Economy," *Proceedings of the National Academy of Science* [vol. 99, no. 4] pp. 9266–9741 (July 9, 2002).
3. Cheslog, C., "New Report Outlines the Ecological Footprint of 146 Nations," Redefining Progress Media Release (2002) www.rprogress.org/media/releases/021125_efnations.html.
4. Youngquist, W., *Geodestinies*, National Book Company, Oregon (1997).
5. "International Energy Outlook," US Department of Energy, Energy Information Administration [Report no. DOE/EIA-0484] (March 26, 2002).
6. McCluney, R., *Humanity's Environmental Future*, SunPine Press, Cape Canaveral, FL, 2004.
7. Pimentel, D., "How Many People Can the Earth Support?" *Population Press* [vol. 5, no. 3] (March/April 1999).
8. Abernethy, Virginia D., "Population and Environment Assumptions, Interpretation, and Other Reasons for Confusion," *Where Next, Reflections on the Human Future*, Board of Trustees, Royal Botanical Garden, London (2000).

Chapter 13

1. Graves, R., *The Greek Myths* [vols. 1 and 2] Penguin, Hamondsworth (1955).

PART IV

1. Until the 1990s, the entire post-war Belle Epoque's oil supply system consisted of seven major companies, nicknamed the Seven Sisters. Since the 1990s, shrinking reserves and "cartelization" has shrunken these seven to five major companies, the Five Dwarfs, which by company capitalization are, Exxon-Mobil, BP-Amoco, Shell, Chevron-Texaco, and Total-Fina-Elf.

Chapter 14

1. Jenkins, G., *Oil Economists Handbook*, Elsevier Science Publishers (1990).
2. "World Energy Outlook," BP-Amoco (2001 and 2002).

Chapter 15

1. McMaster, R.E. (ed.), *The Reaper Investment Newsletter*, www.thereaper. com/, January 2003, cited by Ron Patterson, in Message No. 34313, on the Energy Resources E-List at http://groups.yahoo.com/group/energyre sources/message/34313.
2. The dotcom boom was one of the most unreal, fantastic explosions of hypothetical "value" and following implosion that the so-called "creative destruction" of modern capitalism can wreak. One example of a now forgotten dotcom called priceline.com, created to sell airline tickets on the Internet, was valued at *more than* the combined value of *all* US airlines. See, Cassidy, J., *Dot.Con*, Penguin Books, London (2003).
3. Galbraith, J.K., *The Great Crash*, Houghton Mifflin (1961).
4. The finance sector high-flyer, C. Fishwick, through 2001–02 had operated "split caps" investment packages for around 50,000 investors, whose total loss in that period was around £3 billion. Mr. Fishwick was given a £6 million payoff when quitting the firm "in disgrace" [BBC Radio 4, and *Guardian*, London (October 2002)].

Chapter 16

1. Offroad and agricultural motor vehicles are extensively, perhaps increasingly used for human and goods transport on roads and tracks. This is the case in China and India, where the current lack of road motor vehicles incites the usage of agricultural vehicles for human transport (passengers riding in towed trailers). Fuel efficiency of these "road vehicles" is very low because of technical reasons, that is, vehicles designed for slow speed off-road, being used for road transport.
2. www.geocities.com/MotorCity/Speedway/4939/carprod.html. This website provides data on worldwide car production (from 1995). The very fast growth rates of car production in several countries (ten to 15 times their population growth rates) is clear. 2001 total production = 40.9 million private cars, or about 112,190 per day, worldwide.
3. Tadashi, M., *The Motorisation of Cargo Transport in Japan*, Japanese Department of Transport, Tokyo (1982); Shimokawa, K., "Japan, The Late Starter who Outpaced her Rivals," in *The Economic and Social Effects of the Spread of Motor Vehicles*, Macmillan, Basingstoke (1987); White Paper on Transport, Japanese Department of Transport (1984).
4. Shimokawa, "Japan, The Late Starter who Outpaced her Rivals."
5. Konn, G. and Okano, Y., *Study of Modern Motor Transport*, Tokyo University Press (1979).
6. www.geocities.com/MotorCity/Speedway/4939/carprod.html.
7. SMMT (UK Society of Motor Manufacturers and Traders) statistics for 2001 show the UK having 28.6 million cars and 3.5 million commercial

vehicles in use. This compares with the UK car ownership figure in 1950 of 2.3 million; 1971 of 12.35 million, and 1981 of 15.63 million.

8. "China's Car Sales Hit One Million for First Time. SHANGHAI, Annual car sales in China have topped the one million mark for the first time as a rising urban middle class crowns the world's fastest growing market for foreign automakers, industry experts said on Monday. An official at the China Association of Automobile Manufacturers told Reuters, 1.02 million cars were sold in China in the first 11 months of this year, representing a stunning 55.4 per cent jump from the same period in 2001," Reuters (December 16, 2002).

9. www.geocities.com/MotorCity/Speedway/4939/carprod.html.

10. *Ibid.*

Chapter 18

1. McKillop, A., "Is a 1929-Style Crash Likely This October 2003?" www.vheadline.com, www.gold-eagle.com and www.fiendbear.com (July/August 2003).

2. US Federal Reserve Chairman, Alan Greenspan, made this claim with reference to rising US natural gas prices and "high priced" oil. See, Romero, S., "Short Supply of Natural Gas Raises Economic Worries," *New York Times* (June 17, 2003).

3. For the US in particular, rising oil import dependence is a highly sensitive subject. In the General Accounting Office report, "Energy Security, Evaluating US Vulnerability to Oil Supply Disruptions and Options for Mitigating Their Effects" (Chapter Report, December 12, 1996, GAO/RCED-97-96), the report noted, "The GAO found that, The US economy realizes hundreds of billions of dollars in benefits annually by using relatively low cost imported oil rather than relying on more expensive domestic sources of energy ... oil shocks impose large but infrequent economic costs that, when annualized, are estimated ... tens of billions of dollars per year ..." No mention was made in this report of such a thing as depletion. For EU states dependent on oil and gas from the North Sea there is clear evidence of increasing depletion, meaning increased dependence, as for the US, on oil from the Middle East, Russia, Central Asia and Africa. See also, Blanchard, Roger D., "The Impact of Declining Major North Sea Oil Fields Upon Norwegian and United Kingdom Oil Production," Department of Chemistry, Northern Kentucky University (periodically updated).

4. The OECD-IEA website posts the following "information" regarding the claimed effects of higher oil prices on low-income countries, "High prices hurt poor countries more than rich," "IEA Underlines Developing Nations Dilemma" (posted since 2000). This site claims the following, "High oil prices affect all oil-importing countries but they hurt developing countries more than others. Developing countries suffer more from an oil price hike, as they are more reliant on their energy-intensive manufacturing sectors to spur economic growth. There are often no alternatives to oil. In developing countries an increase in the oil import bill, as a result of a price hike, can lead to a destabilizing deterioration in the trade balance and feed inflation."

5. The energy-intensity, and especially oil-intensity of world economic output is claimed to have fallen by as much as 15–20 per cent over 1980–2000 (see, "World Energy Outlook," OECD-IEA, Paris, various eds.) This analysis focuses on OECD economies and does not closely analyze regional, national and sectoral variations in energy and oil-intensity of output under regimes of prices change. Such study is needed, notably to trace the global macroeconomic impacts of higher oil and energy prices, and forecast demand through 2003–08.

6. The calculation of "oil coefficients of economic growth" is particularly difficult under conditions of falling demand for any reason (very high prices, embargo/rationing, economic recession, etc.). See also, McKillop, A., "Improving the Quality of Oil Demand Forecasts," *Oil & Arab Cooperation Quarterly* OAPEC, Kuwait [vol. 16, issue 59] (1990).

7. *Ibid.*

8. EIA and IEA publications (such as "World Energy Outlook") utilize composite annual growth rates of world oil demand of 1.7–1.8 per cent for projecting world demand at about 115Mbd by 2020. Addressing the May 2002 G-8, Energy Ministers Summit conference in Chicago with the theme, "Preparing for Oil Shocks," US Energy Secretary Abraham projected world oil demand as attaining 120Mbd by 2020. The difference between a 1.7 and a 1.8 per cent annual rate is over 15Mbd through 17 years.

9. The relation of very strong oil and energy demand growth in Asia-Pacific to growing oil import dependence of the US is discussed in, Salameh, M.G., "Quest for Middle East Oil – the US versus the Asia-Pacific Region," *Energy Policy* [vol. 31, pp. 1081–91] Elsevier Press (2003).

10. Concerning natural gas demand growth rates in Asia, note these forecasts for 2003 against 2002, "Individual growth forecasts for some of the major gas consuming countries are India 19.1 per cent, China 13.6 per cent, Taiwan 9.1 per cent, and South Korea 7 per cent." Bo, K., "The Importance of Gas," (editorial) *PetroMin & Hydrocarbon Asia* (May/June 2003).

11. Through the period 1969–2003, the sharpest falls in average annual economic growth rates for OECD countries took place after 1985. See, Jolley, A., CSES Working Paper no. 5, "A New Era of Economic Growth," CSES, Melbourne University, Australia (1996). Average annual growth rates for the G-7 countries fell from 3.2 per cent in 1968–1979, to 1.4 per cent in 1988–1995.

12. Bird, D., "Energy Matters," Petroleumworld.com (June 5, 2003). US oil demand in May 2003 averaged 20.08Mbd compared with 19.48Mbd in May 2002. Growth at 2.9 per cent for the first five months of 2003 was the highest since the late 1970s.

13. Mabro, R., "Does Oil Price Volatility Matter?" *Monthly Comment*, Oxford International Energy Studies – OIES (June 2001).

14. Periodically covered by Reuters' survey of oil market analysts, including Matthew Robinson, "Shrinking Capacity Threatens OPEC Oil Market Rescue," Reuters, Caracas (October 18, 2002).

15. Adelman's "right price" for oil at US$2.50/bbl in 1972 dollars is lengthily defended by Adelman in various works such as, Adelman, M., *The World*

Petroleum Market, Johns Hopkins Press (various editions) and discussed with reference to "dematerialization" of the economy in, McKillop, A., "On Decoupling," *International Journal of Energy Research* [vol. 14, no. 1] (January 1990).

16. See, for example, Meyer, L., "Why Such a Sharp Slowdown?" [remarks by Federal Reserve governor, Laurence H. Meyer before the New York Association for Business Economics and The Downtown Economists (June 6, 2001)]. Meyer claimed that "sharp rises in oil and natural gas prices intensified the fast economic slowdown and slow crash of equity prices."

17. As a sure sign of increasing oil prices, the *Wall Street Journal* (July 29, 2003) carried an editorial bearing the title, "OPEC – One Purely Evil Cartel." In this vituperative editorial, it was claimed that OPEC states are a "gang of price-fixing oil-rich thug regimes [who] meet to reinforce assorted terrorist-sponsoring tyrants at high cost to world consumers."

18. OECD IEA, "Oil Market Report" (July 11, 2003).

19. Periodically covered by Reuters' survey of oil market analysts, including Robinson, "Shrinking Capacity Threatens OPEC Oil Market Rescue," Reuters, Caracas (October 18, 2002).

PART V

Chapter 19

1. Population growth versus food surpluses. The UN FAO website provides data showing annual food surpluses (world food production relative to theoretical needs calculated on a per capita calorie and protein base, for the world's population). These tables show an approximate 13.5 per cent fall in net food surpluses through 1991–2001. Since 2002/03, this has intensified and world food grain surpluses have shrunken rapidly in part due to climate change, and will be further impacted by rising oil and gas prices.

2. Tehran, October 27, 2002, IRNA, "Deputy Head of Institute for Standards and Industrial Research [ISIR] Mohammed Ali Akhavan said here Saturday that if current energy trends continue, the production of oil will equal its domestic consumption and oil exports could dry up by 2008. Ali Bakhtiari dismisses this date as too early, suggesting about 2011 as the date at which Iran will have no exportable surpluses."

Chapter 20

1. Foran, B. and Poldy, F., "Future Dilemmas," CSIRO Resources Futures Working Paper, CSIRO Sustainable Ecosystems. www.cse.csiro.au/ futuredilemmas and the "Australian Resource Atlas" www.audit.ea. vog.au/anra/agriculture/docs/national/Agriculture_Landscape.html which are products of a long-term, ongoing work by CSIRO natural and physical scientists. See also, Newman, S., "The Growth Lobby and its Absence, The Relationship Between Property Development and Housing Industries and Immigration Policy in Australia and France, 1945–2000

with Projections to 2050," pp. 35–51, www.alphalink.com.au/~smnaesp/ populationspeculation/.

2. Flannery, T., *The Future Eaters*, chapter 27, Reed Books, London (1994).

3. *Ibid.*

4. *Ibid.*, p. 114. The "... remarkable assemblage of large, predatory and predominantly land-based reptiles has no parallel outside Maganesia."

5. *Ibid.*, pp. 92–3.

6. Total land stock is 770 million hectares of 7,700,000 square kilometers. Estimates of population range between 150,000 through 300,000 to 900,000. The estimates of people per square kilometer were derived by dividing these total population estimates into total land stock. This does not reflect true settlement patterns, which were lower or denser according to land fertility, which is highest on the southwest and southeast coasts.

7. Flannery, *The Future Eaters*, pp. 148–9.

8. Falling below the UNCCD rainfall benchmarks defined as above the desertification threshold. Almost all of Australia's annual precipitation (465 mm) is evaporated, leaving the lowest runoff for an inhabited continent in the world, at 52 mm water, compared to the 310 mm annual global average. Most living creatures and humans live in the southern coastal regions, where climate is more temperate and rainfall is higher.

9. Foran and Poldy, "Future Dilemmas."

10. "National Report by Australia on Measures Taken to Support Implementation of the United Nations Convention to Combat Desertification," Commonwealth Intergovernmental Working Group for the UNCCD, p. 3 (April 2002).

11. Zable, G., "Population Growth and Energy Sources," www.dieoff.com/ page199.html.

12. Net productivity is measured by carbon gained through photosynthesis and carbon lost through plant respiration. Net primary productivity averages 0.96 Gt of carbon annually for Australia. Nearly 60 Gt of the total continental carbon is stored as plant biomass (45 per cent) and soil carbon (55 per cent). *Australian National Resources Atlas*, http://audit. ea.gov.au/anra/agriculture/docs/national/Agriculture_Landscape.html.

13. Carbon, organic nitrogen and organic phosphorous have nearly doubled since 1788. Some of this is now counterproductive (*Australian National Resource Atlas*).

14. For the lower two figures: Siedlecky, S. and Wyndham, D., *Populate and Perish, Australian Women Fight for Birth Control*, Allen and Unwin [1990], p. 142. For the higher figure: *The Australian People, an Encyclopedia of the Nation, its People and Their Origins*, Angus & Robertson (1988), p. 148 (the higher figure refers to estimates by Noel Buttlin in 1983). They survive in numbers slightly above that lower estimate, but with a variety of lifestyles, from traditional through to post-industrial. Cook and Joseph Banks commented on the low density of the Aboriginal population and Banks correctly inferred this to be a consequence of the low fertility of the land. Thomas Malthus wrote that he was inspired by these comments to write his first volume on population (Malthus, T., *An Essay on the*

Principle of Population and a Summary View of the Principle of Population,
Penguin Classics [1985], p. 251).

15. Flannery, *The Future Eaters,* pp. 368–9. Utilizing 20–30 per cent of the carrying capacity of the land, which is the practice of hunter-gatherers and which suits the El niño southern oscillation (ENSO) dominated climate with its variable cycles of drought and floods. (If further explanation needed, it means that hunter gatherers do not use 100 per cent of resources on an annual basis; they adapt to climatic variations, allowing for more and less productive years by sustaining themselves at about 20–30 per cent of what is available on average over much longer periods than one year.)

16. Suffice to flag some major implications, "carrying capacity" needs to take into account population; energy use per capita and total population; waste products, length of time planned for. Simply put, Australia could have a population as large as Bangladesh, living at Bangladeshi standards, but would only survive for a few days or weeks. The population of the Aborigines, around 300,000, probably survived tens of thousands of years.

17. Foran and Poldi, "Future Dilemmas," p. 125.

18. *Ibid.* Such as legume pastures without balancing lime applications and nitrogenous fertilizers.

19. In this case, "carrying capacity" means long-term, sustainable, the best we can manage. A better term is "optimum population."

20. Foran and Poldi, "Future Dilemmas," p. 121.

21. Rough estimates, based on estimated global stocks of coal, rather than Australia's local stocks. Assumes coal would be shared equitably. Values for coal demand were calculated by finding a value based on current coal demand adjusted for economic growth and adding additional demand due to diminishing oil. Oil and coal production data from the US Energy Information Administration; see also, Vaux, G., "A Projection of Future Coal Demand Given Diminishing Oil Supplies," this volume, Chapter 21.

22. Conventional nuclear energy (as opposed to thorium) is projected to run out around 2100 by sources which also anticipate increase in extraction difficulty will be met with a 300-fold increase in available uranium, after which some expect breeder reactors to be brought online. See, e.g., "Supply of Uranium, UIC Nuclear Issues Briefing Paper no. 75," Uranium Information Centre Ltd., Melbourne (August 2002). See also, Walter, A., *America the Powerless, Facing Our Nuclear Energy Dilemma,* Cogito Press p. 56 (1995). Most objections to nuclear energy are not on grounds that it will run out; rather on health and environmental grounds, or couched in terms of efficiency.

23. Australia has capacity to generate enough electricity through wind power to strongly assist transition to smaller population. Capacity to plan for and implement this is constrained due to the influence of the coal lobby and property development and construction industries, institutionalized at a structural level in Australia. Dr. John Coulter costed such a scheme, based on turbines in the Albany wind farm, Western Australia, which is connected to the West Australian grid through a 15km underground 22,000-volt line." Coulter, J., "Size, Cost and Timing of Change," Paper,

National Conference of Sustainable Population Australia, University of Adelaide 2002. Copies contact: smnaesp@alphalink.com.au.

24. *Agriculture Durable et Conservation des Sols*, European Conservation Agriculture Federation, www.ecaf.org.

25. Ronsin, F., *La Population de la France de 1789 à nos Jours*, Seuil, Paris (1997).

26. As well as the migration of rural women to the cities, seeking more freedom, the industrialization of France relied in part on skilled immigration, which affected population size.

27. The standard workhorse of the SNCF are BB locomotives (1960–1970s vintage still in service) and TER *automoteurs* (some 1950s vintage) relying mostly on diesel fuel.

28. France's electricity production tripled from 1973–2000, growing to 540 TWh, about 80 per cent was nuclear-origin by 2000. French power exports have become an important trade item. However, as "French Nuclear Power and the Global Market" (see Chapter 8) indicates, this is riddled with hidden fossil energy subsidies.

29. The nineteenth-century German geographers Christaller and Losch examined natural settlement hierarchies in traditional Europe. Nested, hexagonal service center hierarchies were the dominant feature of pre-industrial Europe; set by distances able to be covered on foot or horseback.

Chapter 21

1. www.eia.doe.gov (December 5, 2002).
2. Hoffert, M.I. et al., *Science* 981 (2002).
3. www.eia.doe.gov (December 5, 2002).
4. Anthracite and bituminous produce 3.255×10^7 J/kg. Lignite and sub-bituminous produce 2.56×10^7 J/kg.
5. Values for coal demand calculated by finding a value based on current coal demand, adjusted for economic growth and adding the additional demand due to diminishing oil. Oil and coal production data derived from the US EIA (which obtains its data from the World Energy Council).
6. *Ibid.*

Chapter 22

1. For a detailed account see http://www.arts.unsw.edu.au/tsw/.
2. *Ibid.*
3. Douthwaite, R., *Short Circuit*, Green Books, Totnes (1996); Schwarz, W. and Schwarz, D., *Living Lightly*, Jon Carpenter, London (1998); Grindheim, B. and Kennedy, D., *Directory of Ecovillages in Europe*, Germany (1999).

Chapter 23

1. Graves, R., *The Greek Myths*, Vols. 1 and 2, Penguin, Harmondsworth (1955).

Index

Compiled by Sue Carlton